Getting Technical

An Introduction to Technical Writing

Miranda M.H. Oliver

Pippin

Suite 232, 85 Ellesmere Road
Scarborough
Ontario M1R 4B9

Canadian Cataloguing in Publication Data

Oliver, Miranda M.H., 1958-
Getting technical: an introduction to technical writing

ISBN 0-88751-057-4

1. Technical writing. 2. Technical writing — Problems, exercises, etc. I. Title.

T11.055 1994 808′.066 C94-930612-6

Designed by John Zehethofer
Typeset by Jay Tee Graphics Ltd.
Printed and bound in Canada by Friesens

ISBN 0-88751-057-4

10 9 8 7 6 5 4 3 2

To the students, past and present, of
the Arba Minch Water Technology Institute
in Ethiopia

PREFACE

My initial interest in writing sprang from work I did as a research assistant at the Ontario Institute for Studies in Education in Toronto in the early eighties, on a study on writing and cognition in children.

In 1989 I went to Ethiopia to teach English at the Arba Minch Water Technology Institute. I was asked to teach a course on technical writing. There was no curriculum in place, nor were there any instructional materials. All I had at my fingertips was a small library full of undergraduate engineering textbooks, plus a handful of engineering professors who possessed some technical documents that were actually written in English. Teaching the course was a challenge, but by the end of the first semester both my students and I had miraculously acquired a certain proficiency in technical writing.

After that first semester I went to England with the intention of buying some appropriate textbooks on technical writing. I was stunned by the lack of materials. Instead of spending a few minutes in an ESL bookstore I spent several weeks in libraries and bookstores, but still returned to Ethiopia empty-handed though armed with the results of a summer of research. When I finally left Ethiopia, I left behind a manual for the instruction of technical writing. Over the last two years that manual has grown into this book.

Many people have helped me with that growth from manual into book. I would particularly like to thank Mike Price, Tong Yen Lee and my father for providing me with much of the technical information contained in the examples in this book.

My thanks also to Ruth Biderman for her editorial assistance, as well as to the anonymous reviewers whose patient guidance was of immeasurable value.

Last, but not least, thanks to my mother for providing me with free, full-time childcare while I took a leave from my regular work in order to complete this book.

Miranda Oliver
Toronto
March, 1994

CONTENTS

INTRODUCTION

GETTING TECHNICAL: An Introduction to Technical Writing is intended for both speakers of English as a second language and native English speakers who want to learn how to communicate technical information to readers clearly and effectively. It has been designed so that it can be used with or without the help of an instructor.

The book is divided into three sections. Each section begins with a brief introduction that may include suggestions for working with the material that follows. Section I, *Basic Skills*, deals with some of the fundamentals of writing, and provides a foundation for Sections II and III. Section II, *Descriptive Writing*, focuses on the more difficult aspects of technical writing—writing definitions, descriptions, explanations and instructions. Section III, *Report Writing*, offers step-by-step instructions on how to write technical reports.

Since individuals studying or working in a wide variety of fields will be using *GETTING TECHNICAL*, examples that do not require a highly specialized background to be understood have been used whenever possible. It is the format of the examples that is important, not the content. However, some examples in the section on reports are, of necessity, a little more complex than those in the first two sections.

Each chapter includes exercises. For those of you who have no-one to check your work, answers to each exercise have been provided in the *Answer Key* at the back of the book. Since no two people write in the same way, the answers given to exercises that involve writing should be seen as samples only. You can compare your answers to the samples to see if they include roughly the same information and follow a similar format.

To get the most out of *GETTING TECHNICAL* you should not limit yourself to the exercises given in the book. Writing skills improve with practice. After working through a chapter you may find it helpful to apply what you have learned to ideas or material you are familiar with.

It is also important to realize that writing is a process that involves revision and editing. A first draft usually has to be rewritten several times to produce the final version. It is helpful to wait a few hours, or even days before reading over a draft and rewriting it. Taking that time will help you to see your writing as your readers will see it.

SECTION 1

Basic Skills

This section has two parts—a chapter on paragraphs and a chapter on style. These chapters have been included to provide a foundation to work with in sections II and III.

The structure and function of the paragraph are no different in technical writing than in most other styles of writing. Therefore, if you have already studied paragraph formation in another context you may decide to skip this chapter. Keep in mind, however, that if you do not know how to organize a paragraph you will find it very difficult to organize an entire report.

The second chapter focuses more specifically on technical writing. As you work through it, you will learn about some of the considerations that affect style in technical writing. When you finish this chapter, you may find it helpful to look through some technical texts written in your own field of work. That way you can see how the principles you have learned are applied in a context that is familiar to you. Try to read something written not too long ago, since preferred writing styles change over time.

CHAPTER ONE

The Paragraph

The paragraph is an important structure since it serves as the basic building block for most forms of writing. A well written paragraph is easy to read; a badly written one can cause confusion. The key to writing a good paragraph is organization.

A paragraph is a group of sentences that function as a unit dealing with one main idea. The subject of one paragraph is usually closely related to the subjects of those that come before and after it, so you need to know how to decide where one paragraph ends and another begins. You also need to know how to organize the information within a paragraph. Two important features of a well organized paragraph are topic sentences and cohesion.

Topic Sentence

To decide what information should go into a paragraph, you must focus on its main point. Writing a topic sentence will help both you and the reader do that. *A topic sentence is a sentence that clearly states the main idea of the paragraph.* It may appear anywhere in the paragraph but is usually included at, or near, the beginning. The rest of the paragraph develops the idea expressed in the topic sentence.

Given a good topic sentence, the reader should be able to predict what the whole paragraph will be about. However, if the idea is not expressed clearly, the reader may not be certain of the writer's meaning. Can you tell what the writer is talking about in the following sentence, or do you have to guess?

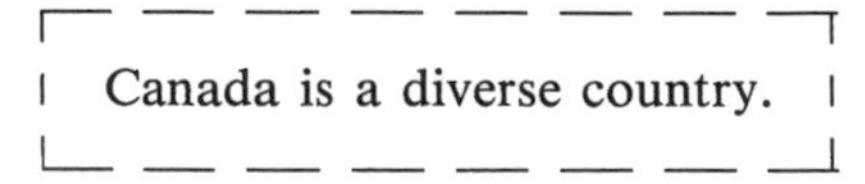

In what way is Canada diverse? The writer could mean many different things. Canada is diverse topographically, ethnographically and climatically. The addition of one of these words would make the meaning clear.

Canada is a climatically diverse country.

Now you do not have to guess what the writer is talking about. Hence, the new sentence could be a good topic sentence.

The definition stresses that a topic sentence *states the main idea of the paragraph.* Since a paragraph should have only one main idea, your topic sentence should not express more than one idea. For example, the following sentence would not be a good topic sentence.

Large cars are expensive to run and they use a lot of fuel.

This sentence contains two different, apparently unrelated ideas:

Large cars are expensive to run.
Large cars use a lot of fuel.

There could be a relationship between those two ideas. If the purpose of the paragraph was to show such a relationship, it could be made into a good topic sentence by rewriting it to make that relationship clear:

* Throughout the book examples are outlined. If the line is broken the example is incomplete, incorrect, or poorly written. If the line is solid, the example is complete and correct.

Large cars are expensive to run *because* they use a lot of fuel.

Now the sentence contains only one idea: the relationship between the cost of running large cars and the amount of fuel they use.

If you were writing a paragraph on this subject, you might think it was obvious that the cost of running a large car is related to the amount of fuel it uses. However, you must not assume that the connection is obvious to your readers. Making false assumptions about a reader's understanding of a subject is one common cause of bad writing. The writer usually knows a lot more about the subject than her readers do, so she must be very careful not to take their understanding for granted.

EXERCISE ONE

Each of the following is the topic sentence for a paragraph. Read each one and predict what the main idea of the paragraph would be.

1. Some materials used in engineering are elastic.
2. Several special tools are needed for electrical work.
3. The procedure for replacing a washer depends on the type of stem assembly involved.
4. Before you take apart a broken dryer, check the simple trouble spots.
5. It is a mistake to tighten the jaws of a pipe wrench too much.

(Answers, p. 183)

Now, look at some examples of topic sentences in context. The following paragraphs come from a text about internal combustion engines. The topic sentences are written in italics. Read the topic sentence for Paragraph 1, then stop and predict what the paragraph will be about. Finish reading the paragraph, and see if your prediction was correct.

1. *In most internal combustion engines, the pistons have to go through a four stroke cycle to produce any power.* The sequence of events in this power cycle includes an intake stroke, a compression stroke, an ignition and power stroke, and an exhaust stroke. Only the ignition and power stroke delivers any actual power. The other three prepare the engine for the power stroke.

The topic sentence informs the reader that the pistons in an internal combustion engine must go through a four-step sequence to pro duce power. The rest of the paragraph outlines the sequence.

The final sentence not only completes Paragraph 1, but it also acts as a bridge to Paragraph 2. It helps the reader make the transition from one idea to the next by preparing her for what is coming. To see how it does this, read the final sentence of paragraph 1 again and then read Paragraph 2.

2. *The power cycle operates in the following way.* First, on the intake stroke, the piston sucks in a mixture of fuel and air through a hole called the intake port. On the compression stroke the fuel-air mixture is compressed to one-eighth of its original volume. This causes the mixture to burn faster and produce more power. When the piston reaches the top an electric spark ignites the fuel-air mixture. This drives the piston back down with a tremendous amount of power. That is why it is known as the power stroke. Finally, as the piston rises, an automatic valve opens the 'exhaust port' and the exhaust gases are pushed out. The cycle can then start all over again.

Paragraph 1 introduces the idea that combustion engines go through a cycle to produce power. Paragraph 2 explains how that power cycle operates. The last sentence in Paragraph 1 is related to both paragraphs, so it acts as a bridge to lead the reader into Paragraph 2.

In Paragraph 2, as in Paragraph 1, the topic sentence was the first sentence. Although that is a common location for the topic sentence, it does not have to be there. In Paragraph 3 the topic sentence comes later. See if you can identify it.

3.	In this cycle the pistons have to go through four strokes to pro duce any power. An engine that only has to go through a two stroke sequence should, theoretically, be twice as efficient. In fact, two-stroke engines are usually not very efficient because of prob lems with their exhaust-and-intake systems. In some cases part of the fuel-air mixture is blown out with the exhaust. In others, some of the exhaust gets trapped in the cylinder, preventing the fuel-air mixture from burning properly. As a result they burn a lot of fuel and can be relatively expensive to operate.

Did you find the topic sentence? It is the third one, which starts with "In fact, two-stroke...". The two sentences before it lead up to the main idea. They tell the reader why two-stroke engines should be more efficient than four-stroke engines. This arrangement helps the reader understand the contrast between four-stroke and two-stroke engines.

The final sentence in Paragraph 3 both completes it and also prepares the reader for the main idea in Paragraph 4. Can you guess what that might be? Read paragraph 4 to see if you are right. Also, try to identify the topic sentence in Paragraph 4.

4.	Automobiles burn a lot of fuel, so fuel economy has to be con sidered when designing a car engine. Therefore, two-stroke engines are never used. Small size and light weight are likely to be more important than fuel economy with machinery that does not use much fuel. That is why two-stroke engines are usually found in power saws, outboard motors for boats, and some lawnmowers.

Were you able to identify Paragraph 4's topic sentence? It is the third one, which begins with "Small size and light weight...".

Now, read the four paragraphs together, paying attention to the role of the topic sentences and the way the final sentence of one paragraph leads you into the next. The topic sentences are written in italics.

In most internal combustion engines, the pistons have to go through a four stroke cycle to produce any power. The sequence of events in this power cycle includes an intake stroke, a compression stroke, an ignition and power stroke, and an exhaust stroke. Only the ignition and power stroke delivers any actual power. The other three prepare the engine for the power stroke.

The power cycle operates in the following way. First, on the intake stroke, the piston sucks in a mixture of fuel and air through a hole called the intake port. On the compression stroke the fuel-air mixture is compressed to one-eighth of its original volume. This causes the mixture to burn faster and produce more power. When the piston reaches the top an electric spark ignites the fuel-air mixture. This drives the piston back down with a tremendous amount of power. That is why it is known as the power stroke. Finally, as the piston rises, an automatic valve opens the 'exhaust port' and the exhaust gases are pushed out. The cycle can then start all over again.

In this cycle the pistons have to go through four strokes to pro duce any power. An engine that only has to go through a two stroke sequence should, theoretically, be twice as efficient. *In fact, two-stroke engines are usually not very efficient because of prob lems with their exhaust-and-intake systems.* In some cases part of the fuel-air mixture is blown out with the exhaust. In others, some of the exhaust gets trapped in the cylinder, preventing the fuel-air mixture from burning properly. As a result they burn a lot of fuel and can be relatively expensive to operate.

Automobiles burn a lot of fuel, so fuel economy has to be con sidered when designing a car engine. Therefore, two-stroke engines are never used. *Small size and light weight are likely to be more important than fuel economy with machinery that does not use much fuel.* That is why two-stroke engines are usually found in power saws, outboard motors for boats, and some lawnmowers.

So far, the topic sentences in the examples have been at the beginning or in the middle of the paragraph. There is one more possibility. A topic sentence can appear at the end, as it does in the following paragraph.

Gasoline evaporates in an open container, leaving behind a gummy residue that can clog small holes. If the holes in a carburetor become blocked by these gummy deposits, the engine can fail. Therefore, only gasoline that has been kept in a sealed container should be used in an engine.

Remember, the purpose of the topic sentence is to state clearly the main idea of the paragraph. It does not matter where it is located as long as it fulfils its purpose. It needs to be there for both the reader and the writer. It helps the reader understand the main point of the paragraph. It helps the writer focus on the main point when she is writing. The reader will probably not consciously identify the topic sentence, but the fact that it is there will help her to understand the paragraph.

EXERCISE TWO

The paragraphs in this exercise have poorly written topic sentences, which are written in italics. Your task is to rewrite the topic sentences. After you have read each paragraph you will find an explanation of the problem and some suggestions to help you.

1. *Civil engineers have to know about piles and there are several ways to classify them.* One is to classify them according to the material from which they are formed. A second method is to examine the way in which they are installed. Another method is to classify them according to their function. Yet another method is to look at the effect the pile has on the soil during installation.

The topic sentence contains two ideas. One is that civil engineers have to know about piles; the other is that there are several ways to classify piles. Write a new topic sentence that contains only one idea. Since the paragraph lists several ways to classify piles, that is the idea that your topic sentence must express.

2. *Piles that are classified according to their effect on soil during installation can be divided into two groups that are classified according to their effect on soil during installation.* The first group is called "displacement piles". This includes piles where the soil is displaced, but not removed, to make room for the pile. The second group is called "non-displacement piles". In this case the soil is removed first in order to create a space for the pile.

The topic sentence is not clear. It is very confusing because it repeats itself. Write a new topic sentence that does not repeat itself.

3. First, we will look at displacement piles. *Some piles are screwed into the ground.* The first group includes piles that are driven into the ground and left in position. They can be solid or hollow, and can be made from a variety of materials. The second group includes concrete piles that are actually formed in place. To form one of these, a pile-like object is inserted into the ground and then withdrawn. The space it leaves is filled with concrete to form the actual pile.

The topic sentence is not related to the rest of the paragraph. The main idea of the paragraph is that there are two different types of displacement piles. Write a new topic sentence that states what the paragraph is about.

(Answers, p. 183)

Paragraph Cohesion

Another way to organize your thoughts and improve your paragraph writing is to make sure that you link ideas together coherently. *A cohesive paragraph is one in which the sentences are organized in a logical way so that the connection between one idea and the next is obvious to the reader.*

The following two paragraphs are both about the same topic. They include exactly the same information and the sentences in each are almost identical. But the first one is not cohesive, whereas the second one is. Read them both to see how much easier it is to understand the second one.

1. While visiting a small, isolated village on the banks of the Nile in Sudan, a German scientist recently discovered that the villagers there have their own answer to the problem of water purification. They add a solution made from the Moringa Oleifana tree to the river water. Conventional treatment of polluted water depends heavily on the use of alum. When the solution the villagers use is added to the river water it must be stirred for five minutes and then allowed to settle for two hours. Alum is prohibitively expensive and can cause health problems if improperly used. Scientists have been searching for an alternative method to be used in the developing world. The solution used by the villagers not only eliminates turbidity but also removes nearly all faecal coliform bacteria. The solution is made by grinding the seeds from the Moringa Oleifana tree and shaking them up with clean water to make a concentrated solution.

2. Conventional treatment of polluted water depends heavily on the use of alum. Unfortunately, alum is prohibitively expensive, and can cause health problems if improperly used. Therefore, scientists have been searching for an alternative method to be used in the developing world. While visiting a small, isolated village on the banks of the Nile in Sudan, a German scientist recently discovered that the villagers there have their own answer to the problem of water purification. They add a mixture made from the Moringa Oleifana tree to the river water. The mixture is made by grinding the seeds from the Moringa Oleifana tree and shaking them up with clean water to make a concentrated solution. The solution is added to the river water, stirred for five minutes, and left to settle for two hours. This process not only eliminates turbidity but also removes nearly all faecal coliform bacteria.

The first paragraph is very hard to follow because it is so poorly organized. It jumps back and forth from one idea to another, and the main point—that the solution used by the villagers in Sudan may be a good alternative to alum—is nearly lost in the confusion. The reader might even be left thinking that scientists are looking for yet a third solution to the water treatment problem.

The second paragraph is much easier to follow. The sentences follow one another in a logical order. First, you are told about the problems related to using alum in conventional forms of water treatment. Then you learn that these problems have led scientists to search for an alternative method that could be used in developing countries. The writer then states that such an alternative has been found in the Sudan, and goes on to say what that alternative is, how it is made, and how it is used—in that order. Finally, you are told how efficient this method is. Each sentence leads into and prepares the reader for the next one. The resulting paragraph is cohesive.

The writer has done two things to make it cohesive. As indicated, she has presented the information in a logical order that the reader can follow easily. She has also used connecting words such as "unfortunately" and "therefore" to indicate the relationship between one idea and the next.

Connecting words like these help to guide the reader. The word "Unfortunately" at the beginning of the second sentence in Paragraph 2 tells the reader she is going to hear about a problem with the information she has just read. The word "therefore", near the beginning of the third sentence, tells the reader that the information in the third sentence is a result of the information in the previous statement. Words like "therefore" and "unfortunately" make the connection between one idea and another obvious. For a list of other connecting words, see Table 1 on page 16.

Here are two more paragraphs that illustrate the difference between an incoherent and a cohesive paragraph.

1. Circular saws are used primarily for cutting lumber. A metal-cutting blade will cut through aluminum, copper, lead or brass. The saw can be fitted with an abrasive disk that can be made of various materials. Each material is suited to a particular task. A friction blade will cut through corrugated iron and thin sheet metal. A disk made of silicon carbide will cut through marble, slate and building blocks. If the correct blade is used, circular saws can be adapted to a number of different tasks. An aluminum oxide disk will cut thin gauge ferrous pipes.

2. Circular saws are used primarily for cutting lumber. However, if the correct blade is used they can be adapted to a number of different tasks. For instance, a metal-cutting blade will cut through aluminum, copper, lead or brass, and a friction blade will cut through corrugated iron and thin sheet metal. Circular saws can also be fitted with abrasive disks made of various materials suited to particular tasks. For example, a disk made of silicon carbide can cut through marble, slate and building blocks. An aluminum oxide disk will cut through thin gauge ferrous pipes.

Once again, the logical way in which the sentences are organized and the use of connecting words make Paragraph 2 cohesive. Their absence in Paragraph 1 makes it very hard to follow what the writer is trying to say. For example, it is not clear whether the second sentence is still talking about circular saws. It appears that the writer may have changed the topic and could be talking about some other kind of cutting device called a metal-cutting blade. In the second paragraph the connections between the sentences are clear.

The following chart gives a brief explanation of the meanings of some connecting words and phrases. If you are not sure how to use them correctly, look them up in a grammar reference book.

although though nevertheless however but in spite of despite	These words and phrases are used to show contrast between two statements. They tell the reader that the second statement will be different than the first statement would lead her to expect.
in fact	This is used to introduce a statement of fact that may oppose or may support the preceding statement.
therefore hence as a result because of this	These words and phrases introduce information that results from the information in the preceding sentence(s).
because	A clause beginning with the word *because* will explain the cause of whatever result is stated in the main clause of that sentence.
also another the other	These words introduce sentences containing more information to support a statement previously made.
first second third next finally	These words can be used for the same purpose as those in the preceding group. They can also be used to indicate the steps in a sequence of events. They let the reader know that she is still reading about one process.
which that who whose when where	Relative pronouns are connecting words that are used to connect two ideas about one subject.

Table 1. Connecting words and phrases

EXERCISE THREE

Read each of the following paragraphs. Rewrite them, making any necessary changes such as adding connecting words or phrases, to make them more cohesive. There is more than one way to revise each one. In each case some suggestions have been made to help you with the rewriting. The sentences in each paragraph are numbered to help you follow those suggestions. Do not number the sentences in your answers.

1. 1.A canister vacuum cleaner has a stronger suction than an upright vacuum cleaner. 2.It has a more powerful motor than an upright. 3.It has detachable parts. 4.The detachable parts make it less cumbersome than the upright for cleaning drapes, window sills and other areas off the floor. 5.An upright vacuum cleaner is more effective for cleaning rugs. 6.It has a spinning beater brush. 7.The beater brush is very effective for loosening and removing the dirt from a rug.

Suggestions:

— Combine sentences 1 and 2.
— Add a word or phrase at the beginning of sentence 3 to show how it supports the idea that a canister vacuum cleaner is better than an upright.
— Combine sentences 3 and 4.
— Introduce sentence 5 with a word or phrase that will prepare the reader for the contrasting idea that an upright vacuum cleaner may be better than a canister.
— Combine sentences 6 and 7.

2. 1.A deep well is required when the water table lies more than 25 feet below the topsoil. 2.A deep well is required when the formations below the water table do not yield water readily. 3.Shale absorbs water. 4.Shale does not yield water. 5.If the formations below the water table are shale a deep well will be needed no matter what the depth of the water table is. 6.If the formations below the water table are shale a shallow well will be unreliable. 7.Drilling a deep well requires professional drilling

equipment. 8.Drilling deep wells is expensive but they are reliable.

Suggestions:

— Combine sentences 1 and 2.

— Add a word or phrase at the beginning of sentence 3 to show that this sentence gives an example of the principle stated in sentence 2.

— Combine sentences 3 and 4.

— Add a word or phrase at the beginning of sentence 5 to show how it is related to sentences 2, 3 and 4.

— Combine sentences 7 and 8.

(Answers, p. 183)

EXERCISE FOUR

The following sentences form a paragraph on the sequence of actions of a gasoline engine, but they are in the wrong order. First, put them in a logical order. Then, combine sentences, add connecting words and make any other changes you think will make the paragraph more cohesive. Sentence 5 is the topic sentence.

1. Within that space, the mixture is fired and the force of the burning fuel turns a crankshaft.
2. The actions include the following.
3. The engine must suck in a mixture of fuel and air.
4. This completes the sequence.
5. All gasoline engines must complete a sequence of four actions to operate properly.
6. This mixture must be squeezed into a small space.
7. Some engines can do this in two strokes, but most engines require four.
8. The gases created by the burning fuel are pushed out of the cylinder into the air.

(Answers, p. 184)

The focus in the last part of this chapter has been on writing cohesive paragraphs. However, cohesion is not only important within the paragraph; the entire text must be cohesive. No matter how well organized individual paragraphs are, if they do not follow each other in a logical way, the resulting text will be incoherent and hard for the reader to follow.

You have seen how to write a cohesive paragraph, and how to link two paragraphs using the concluding sentence of one paragraph as a bridge to the next. With practice, which must include a lot of rewriting, you can learn to write so that your text is cohesive at all levels — within the sentence, within the paragraph, and within the text as a whole.

CHAPTER TWO

Style

In the context of writing, the word *"style"* generally means one of two things. It *may refer to the way or manner in which one writes, or it may refer to the appearance or physical layout of the printed text.* Layout will be covered in the section on report writing. This chapter focuses on manner.

The way or manner in which you write is affected by the purpose of your writing and who your readers will be. It involves making decisions about such things as whether to use the active or passive forms of verbs, whether to use simple or complex sentences, and whether to strive for a formal or informal tone. These and many other factors interact to create different writing styles.

The following style-related issues will be discussed in this chapter:

- complexity of language
- the reader
- formal vs. informal language
- passive and active verbs
- sentence length
- repetition and redundancy
- vagueness

Complexity of Language

In creative writing, people work hard to develop their own distinctive styles. In technical writing, however, you are not trying to impress people with your artistic ability. Your aim is simply to communicate facts and ideas to the reader in a straightforward manner. But there are many different ways to say the same thing. As a technical writer you must choose the clearest, most efficient way. In other words you should *use language that is as straightforward and simple as possible.*

Each of the following three paragraphs gives the reader the same message using different words. Do not worry if you find the first paragraph difficult to understand. The paragraphs that follow are easier, and they each say the same thing. In fact, each successive paragraph is easier to understand than the one before. As you read them, try to see what makes this so.

1. When setting a message down on paper, no matter what the nature of the subject is, it is imperative that the author write his or her ideas in a fashion that will be thoroughly comprehensible to the person reading the writing. In order to succeed at this task, it is necessary that the author know who his readers are likely to be prior to his starting to write. Excessive wordiness is certain to confuse the reader, and will only make a subject that may already be exceedingly difficult to understand more difficult to grasp.

2. When writing about any subject, the author must write in a fashion that will be clearly understood by the reader. To do this, the author must know who his readers are likely to be before he starts writing. Excessive wordiness will confuse the reader, making a difficult subject even more difficult to understand.

3. Authors must write in a way that their readers will understand. To do this, they must know who their readers are likely to be. Excessive wordiness confuses readers.

All three paragraphs have the same number of sentences and say the same thing, but they differ in length and the type of language used. Paragraph 3, the easiest to understand, is the shortest and uses the simplest, most straightforward language.

Now look more closely at the changes that were made in Paragraph 2 to eliminate some of the excessive wordiness from Paragraph 1.

When setting a message down on paper = *When writing*

no matter what the nature of the subject is = *about any subject*

it is imperative that the author write his or her ideas = *the author must write*

in a fashion that will be thoroughly comprehensible to = *in a fashion that will be clearly understood by*

the person reading the writing. = *the reader.*

In order to succeed at this task = *To do this*

it is necessary that the author know who his readers are likely to be = *the author must know who his readers are likely to be*

prior to his starting to write. = *before he starts writing.*

Excessive wordiness is certain to confuse the reader = *Excessive wordiness will confuse the reader*

and will only make a subject that may already be exceedingly difficult to understand more difficult to grasp.' = *making a difficult subject even more difficult to understand.*

If you join all the phrases in bold type together, you get Paragraph 2. Although it is easier to follow than the first paragraph, it still contains unnecessary words. For example, if you say "*Authors* must *write*", you do not need the words *"when writing about any subject"* because you are in effect saying that they must *always* write in a certain way, no matter what they are writing about.

As well, the words *"in a way that their readers will understand"* are easier to follow than *"in a fashion that will be clearly understood by the reader."* One reason for this is the switch from the passive to the active voice. Saying *"will understand"* is shorter and clearer than *"will be understood by"*. This change from the passive to the active form of a verb will be examined in greater depth later in this chapter.

Another reason why *"in a way that their readers will understand"* is easier to follow is that the word *"fashion"* has been replaced by the simpler word *"way"*. You may also find it helpful to note how the change to the plural *"authors"*—and the accompanying *"their"*—is one way to avoid saying *"his or her"* when trying to include both males and females. Another way to use inclusive language is to use singular nouns and switch from the masculine *"him"* and *"his"* to the feminine *"her"* and *"hers"* in different sections of your writing, as done in this book.

There are two more changes in Paragraph 3 that eliminate unnecessary words used in Paragraph 2. It is not necessary to add that the author must know who his readers are likely to be *"before he starts writing."* Saying *"who his readers are likely to be"* implies that he knows this before he starts writing.

Finally, excessive wordiness will confuse readers even if the subject is very simple, so it does not make sense to limit that statement to situations where the subject is difficult to understand. For this reason, that part of the last sentence has been dropped from Paragraph 3.

Paragraph 3 states clearly that excessive wordiness confuses readers. Using unnecessarily big words, or using several words when a few would do, makes it hard for readers to understand your writing. Your job as a writer is to communicate with them, not to confuse, frustrate or annoy them. People reading technical reports or manuals want to get the message quickly and easily. Simple, straightforward language is both easier to read and easier to write.

The Reader

Paragraph 3 also states that an author must know who his readers are likely to be. *Whom you are writing for affects how you should write.* For instance, the language you use and the details you include when describing a carburetor will depend on whom your readers are likely to be. If you are writing for auto mechanics, you can assume that they already know a lot about car engines. Therefore, you can and should use technical language. If you are writing for students of auto mechanics, you should still use technical language, but you will have to add some explanations. If, however, you are writing for customers with broken carburetors, you will have to use simpler language and tell them only what they need to know to have a basic understanding of the problem.

EXERCISE ONE

Read the following three examples of writing about carburetors, and match each one up with one of the groups of likely readers listed here.

Auto mechanics students
Car owners with broken carburetors
Children

1. A carburetor is an important part of a car engine. In order for the gasoline to burn properly it must be combined with air and delivered to the cylinders in a fine spray. The function of the carburetor is to mix the fuel and air in the correct proportions.

2. Most cars run on gasoline. To work properly, the gasoline has to be mixed with air. That is why car engines have carburetors. A carburetor is the part of a car engine that mixes gasoline with air.

3. A fixed-venturi carburetor consists of six systems.
    1. Float system – This system maintains a small quantity of fuel which is ready to be delivered into the air stream.
    2. Idle system – This supplies fuel when the throttle valve is almost or completely closed.
    3. Main-metering system – This supplies fuel when the throttle is open.
    4. Power system – This enables a richer fuel-air mixture to be delivered to the engine when the throttle is fully opened.

5. Accelerator-pump system – This provides the richer fuel-air mixture needed for acceleration.
6. Choke system – This provides the additional fuel required by a cold engine.

(Answers, p. 185)

Formal Versus Informal Language

Writing in a simple, straightforward style does not mean writing in an informal manner. In general, *technical writing is quite formal.* Some of the factors that affect formality are the use of *slang*, the use of *contractions*, and the choice of *the active or passive voice of verbs.*

Slang

Slang is informal language that uses words and expressions whose meanings and usage may change over time. Because it is informal and because its current meaning is not always clear, slang should not be used in formal writing.

Take the word "cool", for instance. It could be defined as a temperature closer to cold than to warm. However, in slang it has a different meaning. Until recently, the slang use of "cool" tended to refer to something that was interesting or excellent. Currently that slang meaning of "cool" is changing, and "cool" now often refers to something that is "uncool". Ten years from now, the slang use of "cool" may have yet another meaning, so it would be foolish to use it in writing intended to communicate information clearly. Also, it sounds unprofessional. Good dictionaries include definitions of the slang meanings of a word. Refer to one for help in avoiding the inappropriate use of such words.

The following paragraph contains two examples of slang. Can you find them?

> Although the Giant Motors Corporation has a world-wide reputation for producing high quality vehicles, the particular model that's been recommended for the Jinka project is a bit of a rip-off. For the same amount of dough, Fieldwagon Inc. offers a model with a far more powerful engine which, nevertheless, has the added benefit of higher fuel efficiency.

"Rip-off" and "dough" are examples of slang. A rip-off is something that is not worth its cost, and dough is slang for money. Such words should not be used in a technical text.

EXERCISE TWO

Each of the following sentences contains an italicised slang word or phrase. If you do not know its slang meaning, predict what you think it might be. Then look it up in the dictionary and replace it with a more formal word.

1. It is imperative to wear protective *threads* at all times when working in the laboratory.
2. Hardened glass lenses cost only a few *bucks* more than regular plastic lenses.
3. It is illegal to drive when *loaded.*

(Answers, p. 185)

Contractions

Another problem in the paragraph on the Giant Motors Corporation is the use of the contraction "that's". *A contraction is an abbreviation formed by combining two words and replacing the missing letters of the second word with an apostrophe.* For example, the contraction of does not is doesn't, of can not is can't, of they have is they've, and of would have is would've. Contractions are used in spoken English and informal writing, but they should not be used in formal writing. Therefore, "that's" should be written out in full as "that has".

EXERCISE THREE

Rewrite the following without any contractions.

> Plastics don't occur naturally. They must be manufactured. A few've been made by modifying natural substances such as cellulose, but the majority of them're made by a chemical process called polymerization. Polythene's an example of a plastic produced by this method. It's a white waxy solid obtained by polymerizing ethylene gas (C_2H_4). Polythene's characteristics include...

(Answers, p. 185)

Passive Versus Active

Consider these two sentences:

> The technician weighed the sample.
> The sample was weighed by the technician.

In the first sentence, the subject "technician" is performing the action of weighing, so the active voice or form of the verb "weigh" is being used. In the second sentence, the subject "sample" is being acted upon, so the passive voice or form of "weigh" is being used.

In the past, the passive form or voice of the verb was used extensively in technical writing. These days it is used less. Using the passive can make your writing sound very formal, but it can also make the sentence structure more complex and harder to follow. For example, compare the passive and active versions of the following sentence.

> An adverse effect is had on filter seal and gasket sections by some fire-resistant fluids. (passive)

> Some fire-resistant fluids have an adverse effect on filter seal and gasket sections. (active)

Both sentences say the same thing, but the second sentence is easier to follow. Here is another example:

> Next, a bridge is to be made from a piece of scrap metal. (passive)

> Next, make a bridge from a piece of scrap metal. (active)

Once again, the active sentence is easier to follow than the passive version. For that reason, it is often preferred. However, there are times when the passive voice has advantages over the active. For example, the passive can be used to avoid using personal pronouns or naming people, as in the following example:

> After collecting the first set of data, I decided to try an alternative method. (active)

> After the first set of data was collected, a decision was made to try an alternative method. (passive)

In the past technical writers would have invariably written the second sentence in order to avoid saying "I". Recently, it has become more acceptable to use personal pronouns, so you might choose to use the active form in this case because it is easier to follow. However, look what happens if you use names in that sentence.

> After collecting the first set of data, Dr. Smith's research assistant, Jan Rogers, decided to try an alternative method. (active)

In this sentence the emphasis seems to be more on *who* made the decision than on what was done. If that is where the emphasis should be, then there is no problem. If, however, the emphasis should be on the choice to try an alternative, the passive version is the better sentence.

There are other reasons why you might want to use the passive. The writer did not mind identifying himself as the decision maker in the sentence, "After collecting the first set of data, I decided to try an alternative method." However, there are times when you do not want to identify someone. This may be a matter of etiquette, as in the following example:

> Dr. Smith made a serious error and destroyed the sample. (active)

In this case, unless you were writing a detailed report on who caused what, you could avoid telling the whole world who made the mistake by using the passive voice.

> A serious error was made and the sample was destroyed. (passive)

You might also want to use the passive when there is no obvious subject for a sentence in the active form. Look at the following sentence:

> Mirrors must be designed to collect the maximum amount of energy from the sun. (passive)

To write it in the active form you would have to find a subject. Several words are possible but none of them sounds quite right.

> *Someone* must design mirrors to collect the maximum amount of energy from the sun. (active)

Scientists must design mirrors to collect the maximum amount of energy from the sun. (active)

There are other possibilities, but it would be difficult to find one that was appropriate. In this case, therefore, it makes more sense to use the passive form.

There are no simple rules that tell you when to use the passive form of the verb. *A very general guideline is that you should only use the passive if using the active will result in an awkward or inappropriate statement.* You will have to use your judgement as to what is awkward or inappropriate.

EXERCISE FOUR

Here are five pairs of sentences. In each case, one version uses the passive voice, and the other uses the active. Choose what you think is the better sentence of the pair, and give a reason for doing so.

1 a. It will be made possible by the use of Maglev trains to cut down on power waste, improving efficiency.
  b. Maglev trains will make it possible to cut down on power waste, improving efficiency.
2 a. People store high-level waste in vaults near reactors.
  b. High-level waste is stored in vaults near reactors.
3 a. Aluminum batteries rely on a reaction between oxygen and aluminum to generate electricity.
  b. A reaction between oxygen and aluminum is relied on to generate electricity with aluminum batteries.
4 a. The lower stopper plate must be slid in and it must be secured with retaining clips.
  b. Slide in the lower stopper plate and secure it with retaining clips.
5 a. It was discovered that the molecule did not behave as expected.
  b. One of the members of our research team discovered that the molecule did not behave as expected.

(Answers, p. 185)

Sentence Length

Style is also affected by sentence length. Ideas can be combined into sentences in many different ways. It is possible to write a grammatically correct sentence that is more than a page long. However, the reader would probably have difficulty understanding it.

The three paragraphs that follow show how sentence length can affect comprehensibility. Read all three and decide which one is the easiest and which one is the hardest to understand. Then, go back and count the number of sentences in each. Can you draw any conclusions about how sentence length affects comprehensibility?

1. The Super-View control panel has indicator lights. It has menu windows. It also has keys. Finally, it has a unique slide switch. This switch enables you to easily control all of the 2500BX's printer functions and features. Among other things it lets you change print color. It lets you change pitch. It lets you change font. It also lets you make fine adjustments to the top-of-form without any difficulty. Perhaps its biggest advantage is that it allows you to check the printer's default control panel selections in a matter of seconds. This is done by sliding the switch through all of its available functions.

2. The Super-View control panel consists of indicator lights, menu windows, keys and a unique slide switch that enables you to easily control all of the 2500BX's printer functions and features such as changing print color, pitch and font and making fine adjustments to the top-of-form without any difficulty, but perhaps its biggest advantage is that it allows you to check the printer's default control panel selections in a matter of seconds by sliding the switch through all of its available functions.

3. The Super-View control panel consists of indicator lights, menu windows, keys and a unique slide switch. This switch enables you to easily control all of the 2500BX's printer functions and features. Among other things it lets you change print color, pitch, and font. It also lets you make fine adjustments to the top-of-form without any difficulty. Perhaps its biggest advantage is that it allows you to check the printer's default control panel selections in a matter of seconds by sliding the switch through all of its available functions.

Most English speakers would find Paragraph 3 the easiest to follow, Paragraph 1 to be slightly more difficult, and Paragraph 2 to be the hardest. If you counted the sentences, you know that the first paragraph had the most sentences, the second one had the least, and the third had a number in between.

Paragraph 2 is actually one long sentence. Although it is grammatically correct, it is very hard to follow. It is called a run-on sentence because it runs on, and on and on. It contains too many ideas, which makes it difficult for the reader to keep track of everything that has been said and how it is all related.

Paragraph 1 contains many short sentences. It is easier to follow than Paragraph 2 but it is not cohesive. With so many short sentences, it is more like a list than a paragraph about one main idea.

Paragraph 3 is the best. Its sentences vary in length, and none of them is so long that the reader might lose track of what is being said. Nor has it been chopped into so many small parts that it reads like a list.

The most important rule about sentence length is that it should vary. When rereading your work, if you find that most of the sentences are very long, see if you can break them up into separate, shorter ones. Likewise, if you find that almost all your sentences are very short, try to combine some of them into longer sentences that still make sense.

The complexity of the idea you are writing about can affect the length of your sentences. Complex ideas expressed in complex sentences can be difficult for the reader to follow. Generally, therefore, you should try to state complex ideas in shorter, simpler sentences. On the other hand, if an idea is easy to grasp then using a more complex sentence structure should not confuse the reader. Here again, however, you will have to use your own judgement.

EXERCISE FIVE

The following paragraph describes the steps involved in removing an outboard motor from a boat. It has been written as a single run-on sentence. Your task is to rewrite it so that it forms a cohesive paragraph consisting of several sentences. This time try to use the imperative or command verb form that is often used when giving directions. For example, say "Disconnect the..." instead of saying "You should disconnect...." Each sentence should contain only one or two of the steps involved in the procedure.

> To remove an outboard motor from a boat the fuel line should be disconnected and the motor should be run in neutral until all excess fuel is removed from the carburetor and then you should disconnect the remote control, electric cable and safety chain from the motor before loosening the thumbscrews and tilting the motor to drain off the water, so that, finally, you can lift the motor from the boat.

(Answers, p. 186)

Repetition

Unnecessary repetition reduces the effectiveness of your writing. Saying the same thing twice, even if you use different words to do so, may annoy or confuse the reader. Consider the following example:

> First you must initially make certain that the temperature is no higher than 40°C.

Since *first* has the same meaning as *initially*, the writer has said the same thing twice in one sentence. Here are two ways to improve that sentence.

> First you must make certain that the temperature is no higher than 40°C.
>
> You must initially make certain that the temperature is no higher than 40°C.

Another example of the same problem is:

Whenever you enter the room, you must always wear your safety goggles.

The words *whenever* and *always* have the same meaning, so one of them has to go.

Whenever you enter the room, you must wear your safety goggles.

You must always wear your safety goggles when you enter the room.

EXERCISE SIX

The following paragraph contains several unnecessary repetitions. Rewrite it, without the repetitions.

The water turbine, which developed out of the waterwheel, is a descendant of the waterwheel. It was invented to drive machinery in mills and factories. However, at the time when it was invented, it was first designed when steam power was coming into heavy use. For this reason, it was rarely used until the 1870's because of this.

(Answers, p. 186)

Vagueness

Vague writing is unclear or uncertain. Vagueness has no place in technical writing, which should communicate information as clearly and accurately as possible. Read the following example of a vague statement.

When a person develops a fever it is important to take certain precautions before it gets too high or some problems may occur.

After reading this sentence would you know what to do if someone developed a high fever? Would you know how high was "too high"? What kind of problems should you expect? The following sentence is much more informative.

When a person develops a fever it is important to try and cool him down before the fever goes higher than 40°C or he may go into a coma.

Instead of vague phrases such as "take certain precautions", "too high" and "some problems may occur", this sentence gives specific information about what to do and what might happen. Here is another example of vagueness.

When starting a car in a cold climate it is a good idea to let the engine run for a while before driving.

How long should the engine run? How long is "a while"? Is it 30 seconds? A minute? Ten minutes? Does letting it run first even matter? The following revised sentence contains that information.

When starting a car in temperatures below freezing you should let the engine run for at least two minutes before driving to prevent the car from stalling.

The rule to be as specific as possible also applies to terminology. Look at Figure 1:

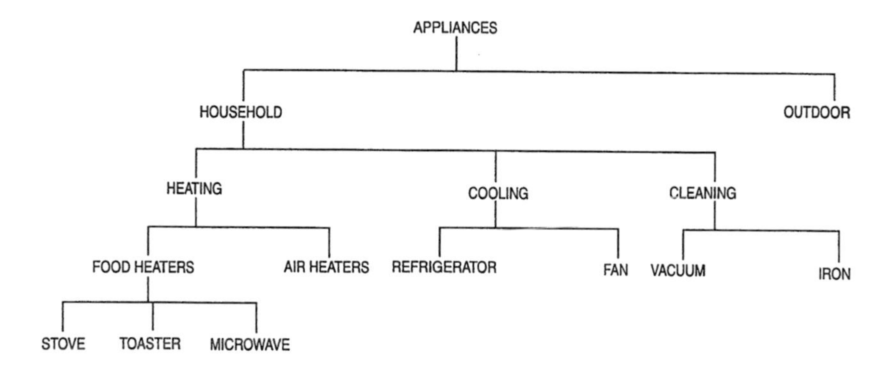

Figure 1. Tree diagram: Appliances

This diagram shows that the term *appliances* can be broken down into categories that can be further subdivided. If you make a statement about *appliances* when you really mean *household heating appliances* you will confuse the reader, as he will think you mean all appliances. Likewise, if you say *heating appliances* when you really mean *microwave ovens*, the reader may conclude that your statement also applies to furnaces, toasters, stoves, and other heating appliances.

But, just as you should not write *heating appliances* when you are only talking about *microwave ovens*, you should not write *microwave ovens* if you are going to talk about many different types of heating appliances. You must be as specific and accurate as possible.

EXERCISE SEVEN

From each of the following pairs, choose the sentence that is more accurate.

1 a. In North America many people use electric appliances to cut the grass.
  b. In North America many people use electric lawn mowers to cut the grass.
2 a. Ford's main business is manufacturing automotive vehicles.
  b. Ford's main business is manufacturing station wagons.
3 a. The oil should be changed every 6 months, or every 12,000 kilometers, whichever comes first.
  b. The oil should be changed at regular intervals.
4 a. Surface wiring can be installed with only a hammer and screwdriver.
  b. Surface wiring, which can be used in many different circumstances, can be easily installed by anyone using a few simple tools.
5 a. A pen is a writing instrument with a small rolling ball in a socket at the tip.
  b. A ballpoint pen is a writing instrument with a small rolling ball at the tip.

(Answers, p. 186)

SECTION 2

Descriptive Writing

Technical writers often have to write definitions, descriptions, explanations and instructions. Writing these well can be challenging, so it is important to work through these chapters carefully, making sure that you have grasped one concept before moving on to the next one. Since an explanation has to include a description and a description has to include a definition, you should do the first three chapters of this section in order.

The first chapter deals with writing definitions. Writing definitions is relatively straightforward and should not give you too much trouble. The next two chapters cover descriptions and explanations. It takes time and patience to write these well, but do not be discouraged. It gets easier with practice.

You will start by describing very simple objects. You should not move on to the section on describing more complex objects until you feel comfortable writing descriptions of simple ones. Do not limit yourself to the exercises given in the chapter; you can practice writing descriptions of any familiar objects. After completing the chapter on descriptions you will learn how to incorporate descriptions into explanations that show how the objects described work.

The final chapter in this section covers instructions. Writing good instructions is easier than writing descriptions, but it still takes practice. One helpful exercise is to write instructions and ask a friend or classmate to follow them. An example of how useful this can be comes from a teacher who once asked his students to give him instructions for erecting the overhead projector screen. They were all familiar with this procedure, but when the teacher followed their instructions exactly, the results were comical. At one point the screen was upside down, at another it was backwards. The students soon saw how instructions can be misinterpreted if not stated clearly. After this exercise, most of the students wrote excellent instructions for erecting the screen.

CHAPTER THREE

Definitions

A definition is a statement that explains what an object is, or what an idea means. As a technical writer you will usually be concerned with writing definitions of objects, so that will be the focus of this chapter.

A good definition should include two types of information:

- a category or classification to which the object belongs
- the object's use or purpose

In each of the following examples the two parts of a definition are given in separate sentences first, and then are combined to form a one-sentence definition.

A distributor is part of the ignition system in a piston engine. (category)
A distributor distributes electric current to the spark plugs. (use)

A distributor is the part of the ignition system in a piston engine that distributes electric current to the spark plugs. (definition)

A chain saw is a power tool. (category)
A chain saw is used for felling trees and cutting logs. (use)

A chain saw is a power tool that is used for felling trees and cutting logs. (definition)

A carburetor is a device in an internal combustion engine. (category)
A carburetor mixes fuel and air in the engine. (use)

A carburetor is a device that mixes fuel and air in an internal combustion engine. (definition)

These examples illustrate the pattern to follow when writing a definition. First, name the object. Second, identify a category to which the object belongs. Third, say how the object is used.

For many people the hardest part of writing a definition is categorizing or classifying the object. One common mistake is to use the word "thing", as in, "A hammer is a *thing* that is used for driving nails into hard surfaces." The word "thing" is too vague. A better definition is, "A hammer is a *tool* that is used for driving nails into hard surfaces."

Table 1 (on pages 41 and 42) lists several words used to categorize objects in technical writing. Many of them have similar meanings. Sometimes, when there is no obvious difference between two terms, one term may be preferred by tradition. In other cases, it may be left to the writer's preference.

TERM	EXPLANATION	EXAMPLE
Machine	An assemblage of parts that work together to perform a specific function.	car, cassette player, fax machine
Device	An assemblage of parts that work together to perform a specific function. It tends to be smaller than a machine. It may be part of a machine.	carburetor, gas meter, bicycle pump
Mechanism	An assemblage of parts that function together to perform an action. A mechanism may be part of a device or machine.	lock, trigger, crane arm
Equipment	A collective noun referring to all the related items needed for a specific activity. In the singular one refers to 'a piece of equipment'.	Construction equipment: backhoe, crane, bulldozer
Apparatus	A set of materials or pieces of equipment that are assembled, often on a temporary basis, to perform a specific function.	scaffold
Structure	Something that has been built. It is used to describe large constructions. It is usually used to describe constructions that are fixed in place.	bridge, building, Eiffel Tower
Construction	Something that has been built.	dam, building
Implement	A simple device or tool that is used to perform a specific type of work.	plough, shovel, crowbar

TERM	EXPLANATION	EXAMPLE
Utensil	A simple device or tool that is used to perform a specific type of work. It is most often used to describe domestic objects.	knife, frying pan, plunger
Tool	An implement that is hand-held. It suggests the need of skill, or a more specific usage than "implement".	wrench, screw-driver, chisel
Instrument	A device or tool used for very delicate or precise work.	scalpel, microscope, calipers
Vehicle	A means of transport usually used to move people over land.	car, wagon
Material	A substance used to make something.	concrete, steel
Object	Something that can be seen or touched. It is solid, not liquid or gas. This word should only be used when there is no more specific word. It is preferable to 'thing'.	

Table 1. Terms for classifying objects

Some objects do not fit neatly into a category, and may be best defined in terms of their shape and what they are made of. For example, a screw cannot easily be defined using any of the terms listed in Table 1, but it could be defined as:

> A screw is a small threaded rod, usually made of brass or steel, that comes to a point at one end. It is commonly used to fasten two pieces of wood or metal together.

Here are two more examples of this type of definition.

A hose is a long flexible tube through which fluids can be moved.

Mercury is a silver-colored element that is commonly used in thermometers.

EXERCISE ONE

On a piece of paper match each word from the list of objects with a word from the list of categories. Then, write a definition for each object. The information in brackets tells you what each object does.

Example: bus — vehicle

A bus is a vehicle used to transport large numbers of people.

Object	*Category*
bus (transports large numbers of people)	implement
magnetic compass (determines direction)	device
sickle (cuts grass)	equipment
pair of water pump pliers (grips pipes)	structure
universal cutter and tool grinder (performs a wide variety of cutting operations)	vehicle
machinist's vise (holds metal or wood securely)	instrument
gear shift (changes gears in a transmission)	machine
tripod (used in surveying)	mechanism
silo (used for storing fodder; is tall)	tool

(Answers, p. 187)

Extended Definitions

An extended definition is one that includes some extra information to help the reader understand what the object is. Many people will get a clearer understanding of a definition if they can visualize the object being defined. Therefore, an extended definition may include an example of a similar looking object. For instance, in the example that follows, the reference to a television screen is a helpful addition to the definition of a computer monitor.

A computer monitor is a piece of equipment that displays the computer output. It resembles a television screen.

An extended definition could also include information about where the object may be found.

An overhead projector is a piece of equipment that is used to project images from a piece of cellophane on to a screen. It is commonly used in classrooms and at conferences.

The additional information in the second sentence may help readers understand the purpose of an overhead projector.

EXERCISE TWO

Rewrite the following definitions as extended definitions. To do this, state what the object resembles or where it is found, or both.

1. A bus is a vehicle that is used to transport large numbers of people.
2. A microwave oven is an appliance that is used for heating foods quickly.
3. A sofa is a piece of furniture that is used to seat several people.
4. A bulldozer is a piece of construction equipment that is used for moving dirt.

(Answers, p. 187)

To decide whether you need an extended definition there are two things you should consider:

Is your definition clear without the extra information? If not, will the extra information make it clearer?

and

For whom are you writing the definition? Are these readers likely to want or need the extra information?

For example, consider this definition:

> The brain is an organ located in the head. It controls and coordinates the body's physical and mental activities.

Would it be helpful to extend this definition by adding the following information?

> It resembles a pile of gray macaroni.

This information might help readers visualize the brain, but it would not help them understand the definition. If anything, it might confuse them by suggesting that there is a relationship between the brain and a type of food. Resemblances should not be mentioned just because they exist, but only if they will make the definition clearer.

The second factor to consider is one that should always influence your writing. Who are your readers and what kind of information do they need? For example, if you are defining a clutch for auto mechanics students, they will need to know what it looks like. How ever, if you are defining a clutch for student drivers, stating what it looks like may not be very helpful at all.

One way to check whether the definitions you write are clear is to leave out the name of the object you are defining and see if a friend or colleague can easily guess what the object is.

CHAPTER FOUR

Descriptions

As mentioned earlier, a definition states what an object is and what it is used for. A description goes one step further. *A description includes a definition plus it tells the reader, in detail, what the object looks like.*

Dimensions

Dimensions are often part of a technical description. If English is your second language you probably know the words you need to express dimensions, such as "length", "width" and "height" but you may have some difficulty using them correctly. For example, when describing a rectangular object, it is grammatically incorrect to say, "It is three meters length", or "It has a wide of two meters." Here are some examples of how to state dimensions correctly.

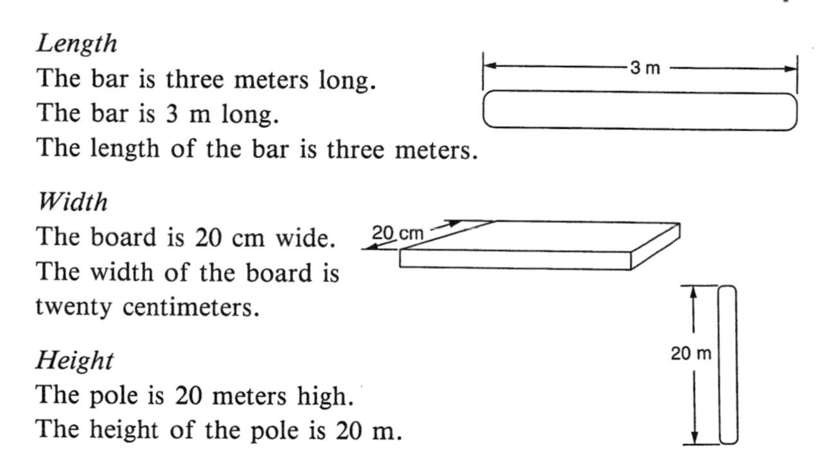

Length
The bar is three meters long.
The bar is 3 m long.
The length of the bar is three meters.

Width
The board is 20 cm wide.
The width of the board is twenty centimeters.

Height
The pole is 20 meters high.
The height of the pole is 20 m.

Depth
The trough is 50 cm deep.
The depth of the trough
is 50 centimeters.

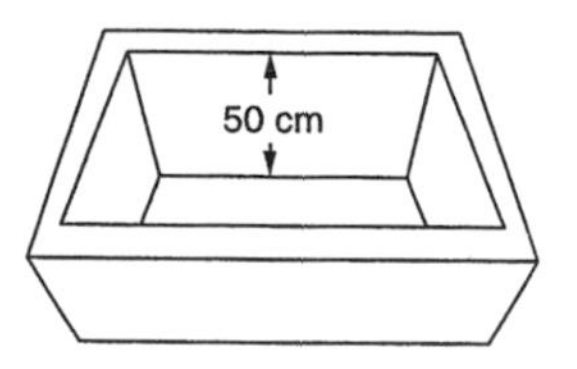

The difference between height and depth can sometimes be confusing. A general rule is that depth is usually used for structures that are below surface level, such as a building's foundation, a ditch or a mine shaft, and for objects which are always approached from above, such as bathtubs or swimming pools. There is another situation where depth may be used. If you are measuring a vertical container that can only be entered from the front, such as a closet or a refrigerator, you may refer to the measurement from front to back as depth. In that case the measurement from side to side would be the width and the measurement from top to bottom would be the height.

Thickness
The board is two centimeters thick.
The thickness of the board is 2 cm.

Thickness is used, rather than width or depth, when the object being measured is solid all the way through. For example, the dimensions used to describe a board would be length x width x thickness.

Diameter
The tube is six centimeters in diameter.
The diameter of the tube is 6 cm.

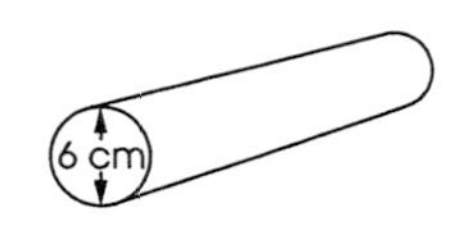

Area (length x width)
The room is 56 m² in area.
The area of the room is
fifty-six square meters.

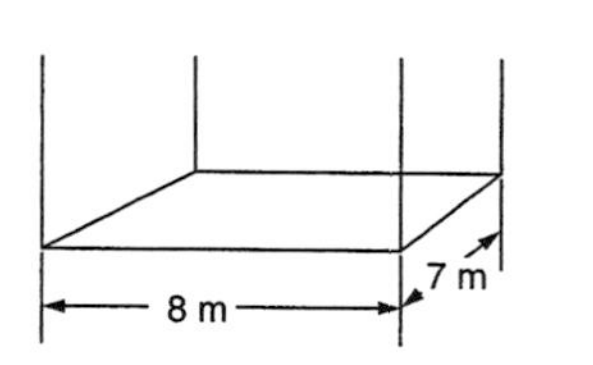

Do not confuse fifty-six square meters (i.e. 7 m x 8 m) with fifty-six meters squared, which would be 3,136 square meters. Fifty-six meters squared would usually be written out as 56 m x 56 m. It is also possible to write it as $(56\ m)^2$, but this practice is less common.

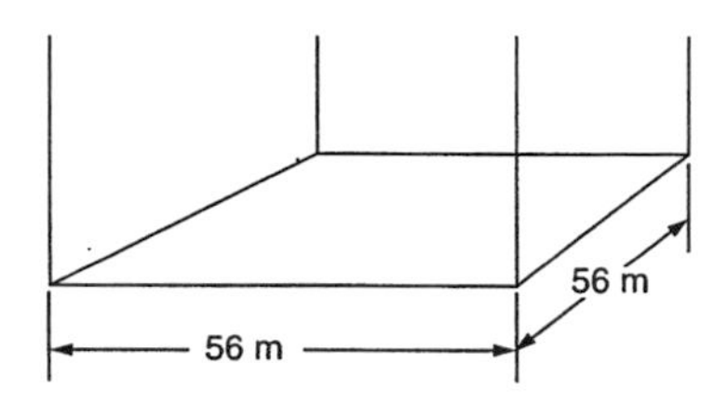

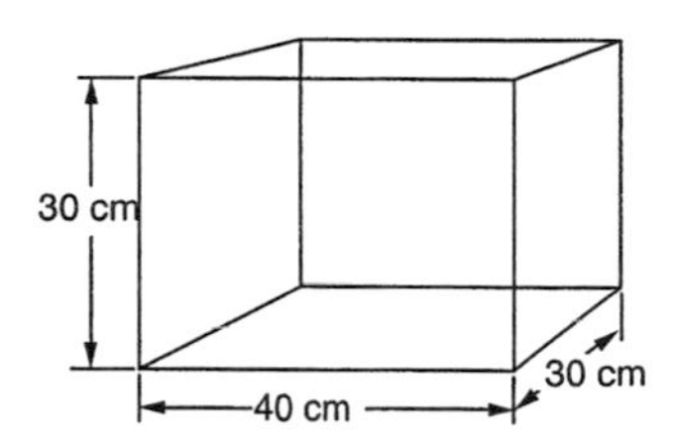

Volume (length x width x height)
The crate is thirty-six thousand cubic centimeters in volume.
The volume of the crate is 36,000 cm^3.

Capacity (length x width x height of interior)
The crate has a capacity of 24,336 cm^3.
The capacity of the crate is twenty-four thousand, three hundred and thirty-six cubic centimeters.
The capacity of the crate is 24,336 cm^3.

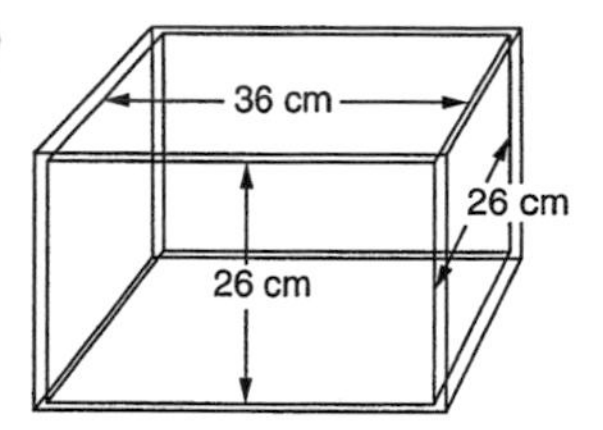

Technically speaking, volume should be used to indicate the total amount of space occupied by the object, whereas capacity should be used to indicate the amount of space contained within the object or structure. In fact, people often do not differentiate between the two and use the term volume for both meanings. People tend to use capacity correctly only when it is important to distinguish between the two measurements.

If the object being measured is used to contain a liquid, one may talk about its liquid capacity. The liquid capacity is the amount of liquid the container can hold.

Weight
The box weighs five kilograms.
The weight of the box is 5 kg.

Density (mass ÷ volume of the object)
The density of the wood is 12 g/cm³.
The wood has a density of twelve grams per cubic centimeter.

Here are some other points to remember when writing about units of measurement.

- Always use internationally approved symbols for units, such as *m* for meter, *km* for kilometer, *L* for liter, *mL* for milliliter, and so on.
- Never pluralize symbols. For example, do not add an *s* to *km* when writing about kilometers.
- Do not use a period after symbols unless, of course, they happen to come at the end of a sentence.
- Always leave a space between the symbol and the numerals associated with it. For example, write 22 g, not 22g.
- Do not use a symbol in a sentence when no numerals are involved. For example, do not write, "Use a m stick to measure off three equal lengths." Instead, write, "Use a meter stick ...."
- All journals, companies and institutions do not use the same style with regard to how numbers should appear in a piece of text. One useful guideline is to write out in full the one-syllable numbers from one to ten, and use numerals such as 11 and 154 for all the rest, as in "You will need one meter of copper wire, and 15 clamps." If you were following this guideline, here is how you would write this sentence, used as an example earlier: "The capacity of the crate is 24,336 cubic centimeters." You would not write, "The capacity of the crate is twenty-four thousand, three hundred and thirty-six cubic centimeters." Another guideline often used is to always use numerals, never words, when dealing with numbers. The important thing for a technical writer to remember is to be consistent throughout any one piece of writing. It is also wise to find out right away what guidelines your company or a new client prefers.

EXERCISE ONE

Each of the following sentences is written incorrectly. Based on the information you have just read about dimensions you should be able to find the errors and rewrite the sentences correctly.

1. The surface is 30 cm long x 20 cm wide. The area of the surface is 600 cm squared.
2. The base is a capacity of 1800 cm^3.
3. The steel must be at least 0.2 mm high.
4. The width of the beam is 11 cm wide.
5. The depth of the building is 50 m.
6. The volume of the cage is 60 cm x 20 cm.
7. The height of the ditch is two meters.
8. The rod is 2.5 cm diameter.
9. The stone is 5 kg weight.
10. The track is 7.8 m in long.

(Answers, p. 187)

Shapes

The following chart lists most of the shapes you will need for writing descriptions in both their noun and adjective forms.

Noun		*Adjective*
square		square
rectangle		rectangular
rhomboid*		rhomboidal
triangle		triangular
circle		circular/round
arc		arcus
oval		ovoid
cube		cubical
sphere		spherical
cone		conical
tube		tubular

* (a rhomboid with equal sides may be called diamond-shaped)

Table 1. Shapes: Noun and adjective forms

Descriptions

There are two types of descriptions.

- General descriptions describe a category of objects, such as cars.
- Specific descriptions describe one specific item, such as a particular model of car.

Both types are examined here. You will first see how to describe simple objects that do not have many parts. Then you will learn to describe more complicated items.

Simple Objects: General Descriptions

There are four steps involved in writing a general description.

1. Write a clear definition.
A general description should always start with a good definition. There is no point in telling the reader what something looks like if he does not know what it is.

2. State the shape, if it was not included in the definition.
Descriptions of simple objects sometimes state the shape as part of the definition, as in, "A can is a *cylindrical* container that is used...." If the shape has not been included in the definition, it should be stated next.

3. State what material(s) it can be made of, if this was not included in the definition.
The definition will very often state what the object is made of. For example, an earlier definition of a screw said it is "a threaded rod, *usually made of brass or steel....*" If the definition does not state what the object is made of, this is the time to do so. If the item being described can be made of different materials, list the most common ones, as in, "Cans are *usually made of steel or aluminum.*"

4. Give typical dimensions for the object.

Here is a general description of a can. The different parts of the description are indicated in brackets.

> (*Definition*) Cans are (*shape*) cylindrical containers that can be used to store and preserve food. They are also used to hold various other substances such as paint or grease. (*Materials*) They are usually made of steel or aluminum. (*Typical dimensions*) Cans come in a variety of sizes. For example, a typical soup can has a liquid capacity of 285 mL and a standard paint can has a liquid capacity of 4.55 L.

EXERCISE TWO

Two general descriptions are given for each of four simple, familiar objects. Use the guidelines you have just been given to select the better of the two descriptions and give reasons for your choice.

1. A brick:
   A. A standard brick is 20 cm long, 10 cm wide and 5.5 cm high. It is a rectangular hardened clay block used in the construction of houses.
   B. A brick is a rectangular hardened clay block used in the construction of houses. A standard brick is 20 cm long, 10 cm wide and 5.5 cm high.
2. A soccer ball:
   A. A soccer ball is an inflatable spherical object used in the game of soccer. It is made of leather with a thin rubber lining. It has a diameter of 22 cm.
   B. A soccer ball is an inflatable spherical object used in the game of soccer. It is usually white and has a diameter of 22 cm.
3. A ballpoint pen:
   A. A ballpoint pen is a cylindrical writing utensil. It consists of a metal or plastic nib attached to one end of a narrow ink-filled tube called a reservoir. The reservoir is encased in a slightly wider plastic tube.
   B. A ballpoint pen is a cylindrical writing utensil. It consists of a metal or plastic nib attached to one end of a narrow ink-filled tube called a reservoir. The reservoir is encased in a slightly wider plastic tube. The average ballpoint pen is 14 cm long with a diameter of 7 mm.

(Answers, p. 188)

The objects described in Exercise Two are common objects. It is unlikely that anyone reading these descriptions would not know approximately how large a pen was, or what a football was made of. However, when you are writing a description of an object that is unfamiliar to your readers these details are very important.

EXERCISE THREE

Write a description of each of the following objects. Use typical or average dimensions.

1. A die
2. A new pencil
3. A jam jar (give either dimensions or liquid capacity)

(Answers, p. 188)

Simple Objects: Specific Descriptions

It is easier to write a specific description than a general description because you are dealing with exact details. The steps to be followed when writing specific descriptions differ somewhat from those used for general descriptions.

1. A specific description should include a narrower definition than a general description.

A general description is written for someone who probably does not know what the object is. A specific description is written for someone familiar with the class of objects, but not with the specific model you are describing. The two definitions that follow illustrate this difference.

> An automobile is a motorized vehicle that is used to transport people along roads. (general)
>
> The Model X is a conventional vehicle with a front-mounted engine and rear wheel drive. (specific)

2. In a specific description you state the shape of the specific object you are describing, which may or may not be typical.

In a general description you state the typical shape of the object being described. For example, most pens are cylindrical, so a general description would say that pens are cylindrical. However, in a specific description you might have to describe a pen that was four-sided or pyramid shaped.

3. In a specific description you state what the specific model you are describing is made of.

In a general description you state what material(s) the object is commonly made of. For example, a table may be made of wood, plastic, glass or metal. In a specific description, you must be as specific as possible, saying, for example, that a particular table is made of pine wood.

4. In a specific description you can give exact, not just typical dimensions.

EXERCISE FOUR

Write a description of the solid oak end table illustrated in the diagram below.

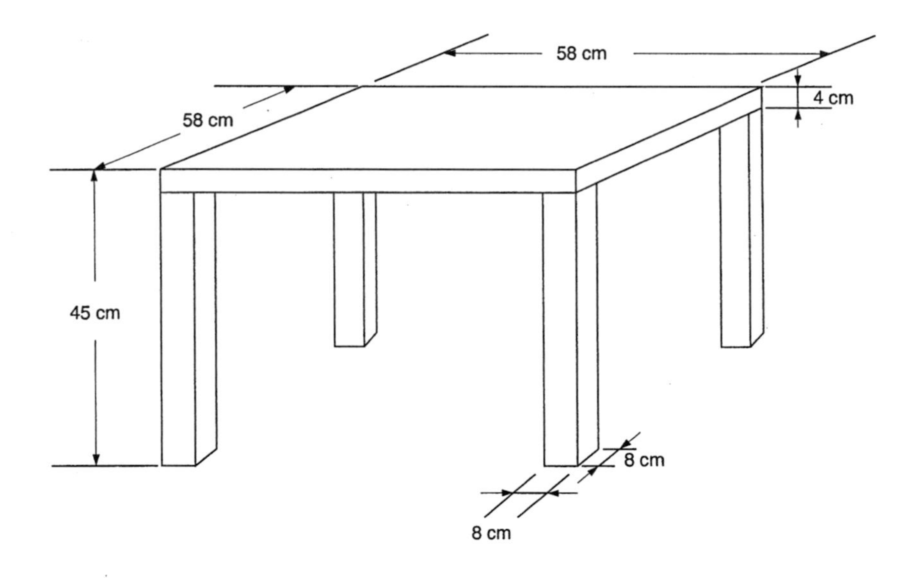

(Answers, p. 188)

Complex Objects: General Descriptions

The procedure for writing general descriptions of complex objects is similar to that for simple objects, but the second step is changed and a fifth step is added.

1. Write a clear definition.

This step is exactly the same as the first step in writing a simple description.

2. State the major components of the object.

It can be very difficult to state the shape of a complex object made up of a number of parts that are each a different shape. Therefore, the second step in a complex description is to state what the major components of an object are. However, this is not the place to go into detail. For example, if you are describing a car you will say that it has an engine, but you will not include the engine's shape or components at this time. That information may come in Step 5. You will mention here that the car has four wheels, but you will not state that each wheel has a tire and a hub cap. You will state that the car has a passenger compartment, but you will not describe its components such as the steering wheel, accelerator, clutch and brake pedals, and turn indicators. The purpose of this step is to give the reader an overall view of the object as a whole.

3. State what material(s) the object is made of, if you have not already done so in the definition.

In this step you should state what the object as a whole is typically made of. For example, the different parts of a car may be made of metal, rubber, leather, plastic or cloth. However, you only need to say what the body is typically made of.

4. State the dimensions.

Here you give typical dimensions for the whole object, not for its component parts.

5. Describe the finer details.

What details you include depends on the purpose of your description. For example, if you are writing a description of an automobile for student drivers you might give only a brief overall description of a car, but go into greater detail about the controls in the passenger section. On the other hand, if you are describing a car for students of auto mechanics you might describe the chassis and the engine in considerable detail but say next to nothing about the interior of the passenger compartment.

The following general description of a standard-transmission car, written for student drivers, illustrates these steps.

> (*Step 1*) An automobile is a motorized vehicle that is used to transport people along roads. (*Step 2*) It consists of a body on a chassis, four wheels, and an engine. The engine is located inside the front part of the vehicle, the passenger section is in the middle, and storage space is at the rear. The vehicle rides on four wheels. (*Step 3*) The body is made of steel. (*Step 4*) An average automobile is 450 cm long, 177 cm wide and 140 cm high, with a wheel base of 260 cm.
>
> (*Step 5*) There are one or two doors on each side of the passenger compartment, and windows on the front, back and sides. The large window at the front is called the windshield. Inside the passenger compartment is a front and a back seat. The driver sits in the front left hand seat.
>
> In front of the driver, below the windshield, is the dashboard. In the dashboard are an assortment of gauges and controls, including gauges to measure the speed of the engine, the speed at which the car is moving, and the amount of gas in the gas tank. There are also heating and air circulation controls.
>
> Mounted on a column extending from the dashboard at approximately chest level is the steering wheel. Levers on either side of the steering wheel column control the lights, turn signals, and windshield wipers. On the right hand side of the column is the ignition slot.
>
> At the driver's feet, on the floor below the dashboard, there are three pedals. The pedal on the left is the clutch, the centre pedal is the brake, and the right hand pedal is the accelerator. Coming out of the floor in the front centre of the car is the gear shift lever. Behind the gear shift, immediately to the driver's right, is the hand brake.

Components or separate parts of the object should be described in a logical order, not just at random. There are a number of ways to do this.

Start with the largest part and move to the smallest.
Start with the most important part and move to the least important.
Start at the front and move to the back.
Start on the outside and move to the inside.
Start at the left and move to the right.
Start at the top and move to the bottom.

The pattern you choose will depend on the object you are describing. Before you write a description, look at the object carefully. Imagine you are looking at it for the first time. What order of presentation would be most helpful to you? Choose the most logical sequence.

Examples A. and B. below illustrate the difference between a description presented in a logical order and one that is not.

A. A bicycle is a two-wheeled vehicle that is ridden by one person for transportation or recreational purposes. It consists of a diamond shaped frame with two wheels aligned one in front of the other. Attached to the frame are a seat, handlebars and pedals. Cheaper bicycles have steel frames. More expensive bicycle frames may be made of aluminum, titanium, carbon fibre or a steel alloy. A ten-speed touring bicycle designed for an adult who is 175 cm tall measures 1 m at the wheelbase and has a 60 cm high frame. Its wheels are 68 cm in diameter.

B. The rear derailleur is attached to the rear fork tip. It descends below the free-wheel cluster, which comprises five sprockets. Attached to the bottom bracket is a double chain wheel. The front derailleur is attached to the down tube, just above the bracket. Center-pull caliper brakes consist of an inner and outer brake arm on a mounting bolt, plus the brake-shoe assembly.

Example A begins with a good overall description. A completed description would include the finer details of the bicycle such as the chain wheel, chain, brakes and gears. A very detailed description would go on to describe the less visible parts such as the hubs and the bottom bracket.

In Example B, the information has not been presented in a logical order. The fact that a bicycle is being described is never even mentioned. Instead of starting with a definition and giving a good overall description of the object, the author has begun by describing some of the smaller parts of the object.

What follows is a detailed, step-by-step example of the process of writing a *general description* of a mercury fever thermometer. First, you will see a diagram, Figure 1, followed by some sentences in bold type that provide a lot of information about the thermometer. However, not all of that information belongs in a description. After each sentence is a statement indicating whether or not the sentence is relevant to a description. When you are reading this, keep in mind the purpose of a description. It is to define an object and give the reader a clear idea of what it looks like. It is not meant to explain how or why it works. Such details belong in an explanation.

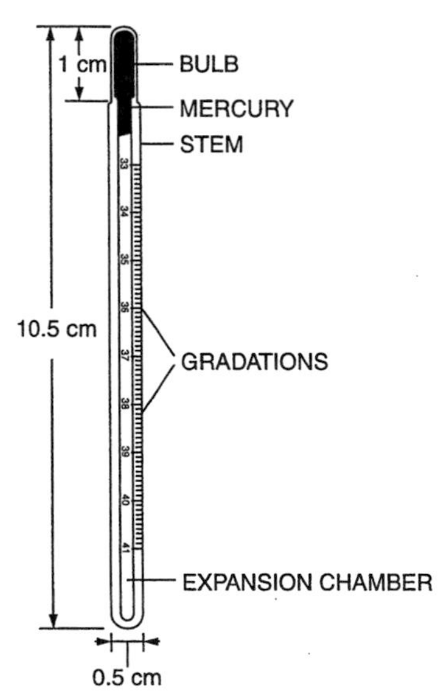

Figure 1. Mercury fever thermometer

1. **The thermometer was developed over three centuries ago.**
   This is an interesting fact but it does not describe the thermometer.

2. **The thermometer is a long glass tube made of two parts:**
   a) a bulb completely filled with mercury.
   b) a stem with temperature gradations marked on it.
   This is relevant to a description. It describes the two main parts of the thermometer.

3. **The mercury in the bulb expands as its temperature rises.**
   This belongs in an explanation rather than a description since it explains how the thermometer works, rather than what it looks like.

4. **To measure a person's body temperature, the bulb end of the thermometer is inserted into the patient's mouth, armpit or rectum.**
   This belongs in an explanation or instructions, as it explains how to use the thermometer rather than describing it.

5. **Inside the stem, fused to the bulb, is a very thin tube called the expansion chamber.**
   This is relevant to a description. It describes part of the thermometer.

6. **Mercury boils at 357°C and freezes at -39°C.**
   This would only be relevant if one were describing a thermometer that had to measure a much greater temperature range.

7. **The expansion chamber contains the mercury in excess of that required to fill the bulb.**
   This is relevant to a description.

8. **Organic liquids such as toluene can be used in thermometers to measure temperatures below the freezing point of mercury.**
   This information would be relevant if one were describing

a broad range of thermometers, but it is not relevant to this description.

9. **The thermometer is used to record the temperature of the human body.**
   This is needed for the definition that comes at the beginning of the description.

10. **Mercury has a high degree of accuracy.**
   This belongs in an explanation, not a description.

11. **The gradation scale on this thermometer extends from 33°C to 41°C.**
   This is relevant to the description.

12. **Mercury expands with increasing temperatures and decreases in volume with decreasing temperatures.**
   This belongs in an explanation, not a description.

13. **Mercury rises up the expansion chamber and indicates the temperature of the bulb on the gradation scale.**
   This belongs in an explanation, not a description.

14. **The body temperature of a healthy human is 37°C.**
   This belongs in instructions, not in a description.

15. **The thermometer is found in homes, hospitals and clinics.**
   This is needed for the definition that comes at the beginning of the description.

Only the information contained in sentences 2, 5, 7, 9, 11 and 15 is relevant to the description. The dimensions of a typical mercury fever thermometer, included with the diagram, are also relevant.

The description can now be written following the steps given earlier in the chapter. The first step is to write a definition. The information for the definition is found in sentences 9 and 15. Putting them together you get:

> The mercury fever thermometer is an instrument that is used to record the temperature of the human body. It is commonly found in homes, hospitals and clinics.

The next step is to state its major components and what it is made of. This information is contained in sentence 2.

> It is a long glass tube consisting of two parts. These are a bulb completely filled with mercury, and a stem marked off in temperature gradations...

The range of the gradations, found in sentence 11, could also be included here.

> ... ranging from 33°C to 41°C.

The dimensions, as shown on the diagram, are included next.

> A typical thermometer is 10.5 cm long and 0.5 cm wide, with the bulb section alone being 1 cm long.

You will notice that the phrase "the bulb section alone" makes it clear that the length of the bulb plus the stem is 10.5 cm, and that the bulb's 1 cm is part of the overall length.

Finally, finer details from sentences 5 and 7 are included.

> Inside the stem, fused to the bulb, is a very thin tube called the expansion chamber. It contains the mercury in excess of that required to fill the bulb.

Here is the completed description.

The mercury fever thermometer is an instrument that is used to record the temperature of the human body. It is commonly found in homes, hospitals and clinics. It is a long glass tube consisting of two parts. These are a bulb completely filled with mercury, and a stem marked off in temperature gradations ranging from 33°C to 41°C. A typical thermometer is 10.5 cm long and 0.5 cm wide, with the bulb section alone being 1 cm long. Inside the stem, fused to the bulb, is a very thin tube called the expansion chamber. It contains the mercury in excess of that required to fill the bulb.

Complex Objects: Specific Descriptions

As with a simple object, a specific description of a complex object is similar to a general description, but must include more information. The definition should state the specific model of the object being described, not simply classify or categorize it. As well, details such as size, shape, and what it is made of, can be exact.

EXERCISE FIVE

The sentences below give you a lot of information about septic tanks. Some of the sentences refer to septic tanks in general. Others refer to a specific model, the SR2. Figures 2 and 3 on pages 65 and 66 illustrate the SR2 from two different angles.

1. Using only the information contained in the sentences that follow, write a general description of septic tanks. Unlike fever thermometers, which generally come in a standard shape and size, septic tanks come in a variety of shapes and sizes. Therefore, the only dimension that your description should give is the capa city of a typical tank. Do not include irrelevant information.
2. Using the information from *both* the sentences *and* from Figures 2 and 3, write a description of the SR2 septic tank. Remember, since this is a specific description your definition will not say what a septic tank is. It will state what kind of septic tank the SR2 model is. Do not include irrelevant information.

   1. The SR2 is a rectangular tank.
   2. Tightly sealed covered manholes provide access to the inlet and outlet devices in a septic tank.
   3. The SR2 has a capacity of 3,400 liters.
   4. A septic tank is a watertight receptacle.
   5. Septic tanks come in a number of shapes. The most common are horizontal or vertical cylindrical tanks and horizontal rectangular tanks.
   6. A septic tank for a four bedroom house should have a capa city of at least 4,500 liters.
   7. Septic tanks commonly have one or two compartments. A tank with one compartment is called a single compartment

tank. A tank with two compartments is called a two compartment tank.

8. The outlet device must retain scum in thc tank.
9. The purpose of a septic tank is to separate solid waste from liquid sewage. The solids are stored in the tank until sufficiently broken down to be discharged for final disposal.
10. Septic tanks are generally made of pre-cast concrete or welded sheet steel.
11. The inlet device must divert the incoming sewage downwards.
12. The SR2 is made entirely of pre-cast concrete.
13. The outlet device may consist of a vented tee or baffle and an outlet pipe.
14. The inlet device may consist of an inlet pipe and a vented tee or baffle.
15. A septic tank should not be closer than 1.5 m to the foundation of a building.
16. In a gravity-feed system, the outlet pipe must be several centimeters lower than the inlet pipe.
17. A septic tank must have an inlet device at one end and an outlet device at the other end.
18. The SR2 is a single compartment tank.

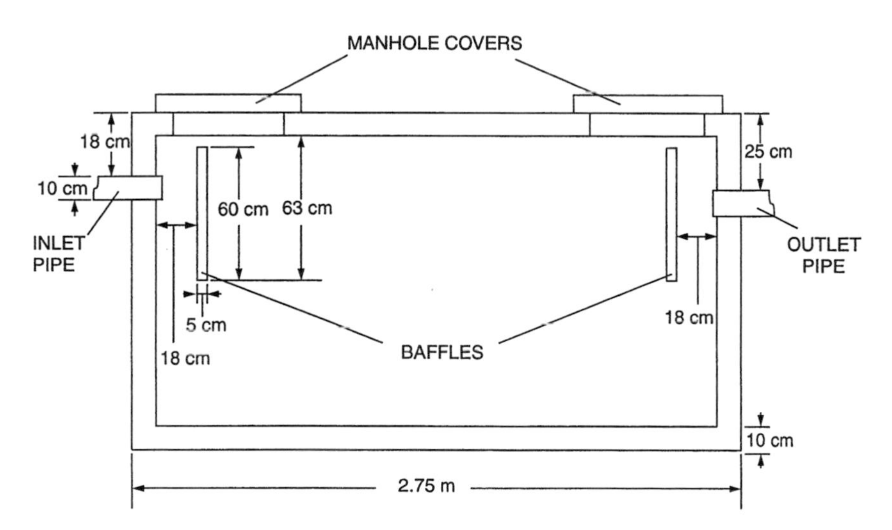

Figure 2. Profile of a septic tank, model SR2, with one side removed.

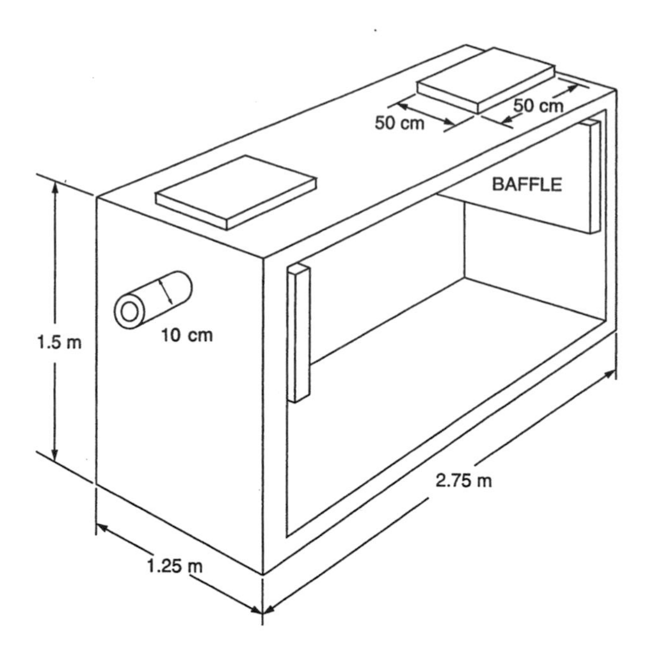

Figure 3. 3-D view of a septic tank, model SR2, with one side removed.

(Answers, p. 188)

It is worth noting here that some objects are almost impossible to describe accurately using words alone. A diagram may be needed too. In some cases, the reader may even use the written description to help him understand the diagram, rather than the other way around.

You do not have to be an artist to draw simple diagrams. The diagrams in this book have all been drawn by the author, who is not a graphic artist. Using a ruler, and possibly a mathematical compass, you should be able to draw diagrams that will be sufficient for most purposes. As well, there are many computer-assisted drawing programs available that can make drawings look very professional.

CHAPTER FIVE

Explanations

Like a description, an explanation defines an object and gives the reader a clear visual image of it. It then goes one step further. *An explanation states how the object works.* There are three steps to be followed when writing an explanation.

1. Define the object.
2. Describe what it looks like.
3. Explain how it works.

The first two steps involved in writing an explanation of a mercury fever thermometer—defining and describing it—were completed in Chapter Four. To carry out the third step you must explain how the thermometer works.

EXERCISE ONE

Chapter Four included fifteen sentences about the mercury fever thermometer. Six of those sentences were used to write a description. The information needed to write an explanation can be found in the remaining sentences, which are given below. Go through these sentences and decide which ones belong in an explanation.

1. The thermometer was developed over three centuries ago.
2. The mercury in the bulb expands as its temperature rises.
3. To measure a person's body temperature, the bulb end of the thermometer is inserted into the patient's mouth, armpit or rectum.

4. Mercury boils at 357°C and freezes at -39°C.
5. Organic liquids such as toluene can be used in thermometers to measure temperatures below the freezing point of mercury.
6. Mercury has a high degree of accuracy.
7. Mercury expands with increasing temperatures and decreases in volume with decreasing temperatures.
8. Mercury rises up the expansion chamber and indicates the temperature of the bulb on the gradation scale.
9. The body temperature of a healthy human is 37°C.

(Answers, p. 189)

Having found the applicable information, the next step is to write the explanation and add it to the description which is repeated below. You may also find it helpful to refer back to the diagram of the mercury fever thermometer on page 60 in Chapter Four.

> The mercury fever thermometer is an instrument that is used to rec ord the temperature of the human body. It is commonly found in homes, hospitals and clinics. It is a long glass tube consisting of two parts. These are a bulb completely filled with mercury, and a stem marked off in temperature gradations ranging from 33°C to 41°C. A typical thermometer is 10.5 cm long and 0.5 cm wide, with the bulb section alone being 1 cm long. Inside the stem, fused to the bulb, is a very thin tube called the expansion chamber. It contains the mercury in excess of that required to fill the bulb.

Having completed Exercise 1, you know that sentences 2, 3, 6, 7, and 8 contain information that should be included in an explanation. To understand how the thermometer is used to measure a person's temperature, you have to understand the qualities of mercury that make the thermometer work. Sentences 6 and 7 explain that, and could lead to this sentence:

> Mercury, which has a high degree of accuracy, expands with increasing temperatures and decreases in volume with decreasing temperatures.

Sentences 2, 3, and 8 explain how the thermometer works and should be included:

> To measure a person's body temperature, the bulb end of the thermometer must be inserted into the patient's mouth, armpit or rectum. The mercury in the bulb expands as its temperature increases. When it expands it rises up the expansion chamber. The level it reaches on the gradation scale indicates the temperature of the mercury in the bulb.

The temperature of the bulb will be that of the person whose temperature is being taken. Therefore sentence 8 can be changed to say:

> It will rise up the expansion chamber and indicate the person's temperature on the gradation scale.

By putting all the parts together we get the following explanation:

> (*Definition*) The mercury fever thermometer is an instrument which is used to record the temperature of the human body. It is found in homes, hospitals and clinics. (*Description*) It is a long glass tube made of two parts. These are a bulb completely filled with mercury, and a stem with temperature gradations on it which range from 33°C to 41°C. The total length of the thermometer is 10.5 cm, with a width of 0.5 cm. The bulb is 1 cm long. Inside the stem, fused to the bulb, is a very thin tube called the expansion chamber. It contains the mercury in excess of that required to fill the bulb.
>
> (*Explanation*) Mercury, which has a high degree of accuracy, expands with increasing temperatures and decreases in volume with decreasing temperatures. To measure a person's body temperature, the bulb end of the thermometer must be inserted into the patient's mouth, armpit or rectum. The mercury in the bulb expands as its temperature rises. It will rise up the expansion chamber and indicate the person's temperature on the gradation scale.

EXERCISE TWO

Read the following explanation of a ballpoint pen. Pick out the part that defines the pen, the part that describes it, and the part that explains how it works.

A ballpoint pen is a cylindrical writing utensil. It consists of a metal or plastic nib attached to one end of a narrow ink-filled tube called a reservoir. The reservoir is encased in a slightly wider plastic tube. The average ballpoint pen is 14 cm long and 7 mm in diameter.

A small ball, which is usually made of steel, is housed in a socket at the tip of the nib. When the tip of the pen is pressed down and moved across a piece of paper, the ball rolls. As it rolls, it transfers ink from the ink tube to the paper.

(Answers, p. 190)

EXERCISE THREE

Using *all* the information from the nine sentences that follow, write an explanation of a mathematical compass. The diagram in Figure 2 is included to help you visualize the compass. You may find it useful to begin by going through the sentences to decide whether they belong in the defining, describing or explaining part of the explanation.

1. One arm ends in a sharp point while the other arm holds a pencil lead at the end.
2. To draw a circle, the arms are opened until the distance between the two outer points is equal to the radius of the circle.
3. The arms of a typical compass are 12 cm long.
4. A mathematical compass is an instrument that is used for drawing circles and arcs.

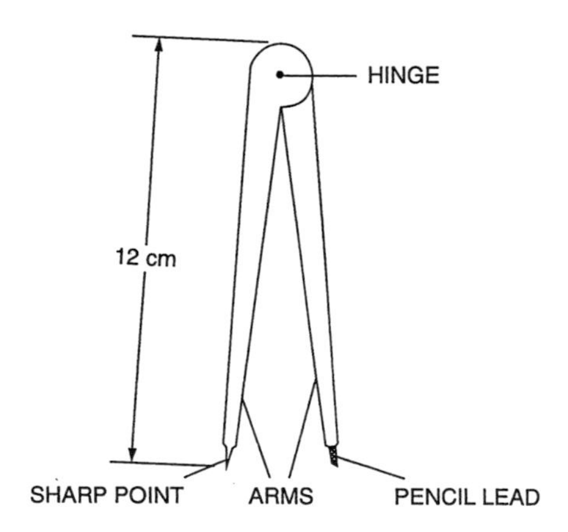

Figure 2. Mathematical compass

5. A compass may be made of metal or plastic.
6. A compass consists of two arms of approximately equal length.
7. The sharp point is placed where the center of the circle will be on a piece of paper.
8. The arms are hinged at one end so that they can be moved together or apart.
9. The sharp point is held in place while the other arm pivots around it with the lead pressed against the paper. In this way, the circle is marked on the paper.

(Answers, p. 190)

Writing explanations of both simple and complex objects involves defining the objects, describing them and explaining how they work. But, when you are writing an explanation about a complex object with many parts that have different functions, you should explain the purpose of the individual parts as you describe them. Then you can explain how the object works as a whole in the final stage.

For example, if you are writing an explanation about a car, when you first mention the clutch you can say that the clutch enables the driver to switch gears. The place where you describe the brakes is the best time to explain that the brakes enable the driver to stop the car. Then, when you have described all the major parts and explained their individual functions, you can finish with an explanation of how the car works as a whole.

Read the following explanation of an overhead projector. The text is divided into its three parts: definition, description and explanation. Within the description the purposes of several parts have been explained. Those phrases or sentences are in italics.

(Definition) An overhead projector is an electrical device that is used to project images from a piece of cellophane called a transparency onto a screen. It is used by teachers and lecturers in classrooms and at conferences. *(Description)* Although the design may vary slightly from one model to the next, a typical overhead projector has three main parts. These are the base, a post, and a reflector mechanism.

The base, which is hollow, is square in shape. It is approximately 36 cm x 36 cm, and is 25 cm high. Its bottom and sides are made of sheet metal. Its top, or cover, is made of glass, and is fastened to the base on one side with hinges *that enable it to be opened.* There is an opaque glass lens fastened to the underside of the glass cover. *This lens diffuses the light so that the image reflected onto the screen is of equal intensity.*

Inside the base there is a small 250 watt light bulb suspended over a concave chromium dish approximately 5 cm in diameter. *The chromium dish deflects the light up through the cover.* A ventilator *to keep the machine cool* is also located in the base.

Attached vertically to one corner of the base is a square post. It may be 2 cm x 2 cm and rise approximately 40 cm above the base. Attached to this post is a steel or aluminum arm that suspends the reflector mechanism above the center of the box. *Turning a knob on the arm raises or lowers this mechanism.*

The reflector mechanism consists of two parts: a convex glass lens and a mirror. The lens has a diameter of approximately 10 cm and rests in a horizontal position. The mirror, about 15 cm x 12 cm, is positioned above the lens, facing downwards. It can be tilted to a maximum of 60 degrees facing front or back. When it is horizontal it is approximately 10 cm above the lens.

(Explanation) To use the projector, a transparency is placed on top of the base. When the machine is turned on the light in the base is projected up through the transparency. The image is thereby projected up through the convex lens and onto the mirror. It is deflected off the mirror and onto a screen. Tilting the mirror to the correct angle enables the user to centre the image on the screen. The focus can be adjusted by moving the reflector mechanism up or down the post as required.

This explanation of an overhead projector was written to illustrate the different parts of an explanation. However, one important part of the explanation is missing. There is no diagram. Just reading the description of the machine is not likely to give you a clear visual image of the object. You can probably picture the base and the post, as they are very simple shapes. However, the reflector mechanism is more complex. It would take many paragraphs to describe it accurately using words alone. If a diagram were included, the explanation above would be sufficient.

One final point needs to be made. Some people make no distinction between descriptions and explanations. Therefore, if you are asked to write a description you should check to see whether a description or an explanation is required.

EXERCISE FOUR

Read the following description of a magnetic compass. Below the description are several sentences containing more information about compasses. Using the description and the relevant information from the sentences below it, write an explanation of a magnetic compass.

> A magnetic compass is an instrument that is used to determine direction. It often looks like a small circular container made of plastic or metal with a clear glass or plastic cover. Inside the container there is a magnetized needle and a compass card that indicates the 32 points of direction. A typical hand-held compass has a diameter of 5 to 10 cm and is 1 to 2 cm deep.

1. The magnetic north pole is very close to the geographic north pole.
2. A magnetic compass must be made of a non-magnetic material such as brass, plastic or aluminum.
3. Gyro compasses can be used to determine true north.
4. By rotating the compass so that north—as indicated on the compass card—is aligned with the needle, you can also see where all the other directions lie.
5. No matter which way a magnetic compass is held, the needle will always swing around so that it is pointing toward the magnetic north pole.

6. The magnetized needle in a magnetic compass is poised on a point in the center of the compass in such a way that it can pivot freely.
7. Gyro compasses are subject to fewer errors than magnetic compasses.

(Answers, p. 190)

CHAPTER SIX

Instructions

Instructions explain how to do something. *Well written instructions give the reader a sequence of steps to follow in order to perform a task successfully.*

Instructions do not explain how or why something works; they only explain how to assemble or operate it. For example, details about the mechanics of an internal combustion engine should not be included in the instructions for operating a car. New owners turn to those instructions to learn how to do such things as start the car and operate the lights, windshield wipers and turn signals.

Knowing who your likely readers are also helps you decide on the most appropriate terminology to use. If, for example, you were writing instructions for the operation of a new water pump, the language you used would be different for a farmer than for an irrigation engineer. As well, you might have to explain more of the terminology for the farmer and go into greater detail in some places because he would probably know less about pumps in general than the irrigation engineer.

Three guidelines for writing instructions are:

1. Instructions must be as simple as possible.
2. Instructions must be complete.
3. Instructions must be clear.

Instructions must be simple.

Instructions, like other technical writing, must be as straightforward as possible. Consider these two examples of the same instruction.

A. The first step in the process involves warming up the equipment before doing anything else. This is easily achieved by simply rotating the on/off switch 30° in a clockwise direction from its starting place so that the arrow on the knob is aligned with the warm-up arrow on the control panel.

B. To warm up the equipment, turn the on/off switch approximately 30° clockwise to the warm-up position marked on the control panel.

Example B is obviously the better instruction. It is clear, simple and concise. Here are two more pairs of instructions. In each case, Example B is better than Example A. Study each pair carefully to see how improvements are made.

A. Coat each wire with lubricant. When you do this, make sure that the entire surface of each individual wire is thoroughly coated.

B. Thoroughly coat the entire surface of each individual wire with lubricant.

A. It is necessary to use a solder that is silver-based. This solder must have a melting point of 850°C.

B. Use a silver-based solder with a melting point of 850°C.

EXERCISE ONE

The following four instructions are written in an unnecessarily complicated manner. Rewrite them so that they are simpler and clearer.

1. Get a beaker and get some water and pour some of the water into the beaker until there is 400 cm^3 of water in the beaker.
2. Get some powdered alum and measure out 80 grams of it and then add the 80 grams to the water in the beaker.
3. Take the beaker with the water and powdered alum in it and place it in a can that is half full of water.
4. The can with the water and the beaker in it must be heated for two to three minutes, so you should place it over a bunsen burner and heat it.

(Answers, p. 191)

Instructions must be complete.

This guideline sounds easy to follow, but a very common problem with written instructions is that they are incomplete. When you know how to operate a piece of equipment you perform many of the steps without being consciously aware of them. When you have to write down these steps, you may forget to include some of them. A reader who is not familiar with the equipment will not notice that you have left anything out, and may very well fail to complete the task you explained. If he does not fail, he will at least be frustrated as he struggles to follow your incomplete instructions.

Imagine you are attempting for the first time to do some work on your car, and you come across this instruction: "Remove the battery." If you are not familiar with automobiles you may not know how to remove the battery. Complete instructions will explain, step by step, how to remove it.

1. Loosen the clamp retaining nuts and bolts to disconnect the leads from the battery terminals.
2. Unscrew the clamp bar retaining nuts, then remove the clamp.
3. Lift the battery from the carrier, keeping it vertical to ensure no electrolyte is spilled.

These instructions should include a diagram that labels the parts.

EXERCISE TWO

Here are two sets of instructions (A and B) on how to start and begin driving a car with manual transmission. These instructions are intended for someone who has never driven before. Each set is incomplete. Read the first set (A) and think about some of the steps that may have been omitted. Do the same thing with the second set (B). Then, using information from both A and B, write a complete set of instructions.

A. 1. Fasten your seat belt as soon as you get in the car.
2. Put the key in the ignition slot. Turn the key clockwise as far as it will go, while pressing on the accelerator pedal until the engine starts. Release the key. If the engine falters press down again on the accelerator pedal.
3. Move your foot onto the brake pedal, press it down and hold it there.
4. Keeping your foot on the brake, depress the clutch pedal with your other foot. While the clutch is depressed move the gear shift into first gear, as shown on the gear shift handle.
5. Move your foot onto the accelerator pedal and slowly push down on it while gently releasing the clutch pedal. The car will start to move forward.

B. 1. Make sure the car is in neutral gear. If it is not, depress the clutch pedal with your left foot and move the gear shift into neutral position, as shown on the gear shift handle.
2. Make sure the hand brake is on. The lever must be up, not horizontal.

3. Put the key in the ignition slot. Turn the key clockwise as far as it will go, while pressing on the accelerator pedal with your right foot. If thc cnginc faltcrs prcss down again on thc accelerator pedal.
4. Move your right foot onto the brake pedal.
5. Release the hand brake.
6. Depress the clutch pedal with your left foot. While it is depressed move the gear shift into first gear, as shown on the gear shift handle.
7. Move your right foot onto the accelerator pedal and slowly push down on it. The car will start to move forward.

(Answers, p. 191)

Instructions must be clear.

Written instructions are often unclear for the same reason that they may be incomplete. The writer may think his instructions are easy to understand because he already knows what to do.

The following all have a part to play in ensuring that instructions are written clearly:

- title
- required tools and materials
- sequence
- layout
- language use
- diagrams

Title

Make sure your instructions have a title that states their purpose clearly. For example, instructions for building a stone retaining wall should not be given a vague title like this:

Retaining Walls

A far better title would be:

How to Build a Stone Retaining Wall

EXERCISE THREE

The instructions in Figure 1 will be rewritten several times over the course of this chapter. Because they do not have a title, the reader cannot tell exactly what they are for until the very end. Your first task is to read them and write a title for them.

(Answer, p. 191)

Using two tire levers, remove the inner tube from the wheel. To do this, you will first have to remove the wheel from the frame. To do this, loosen the bolts on either side of the wheel hub using two wrenches.

Next, pump up the inner tube. Then, place your hands about 15 cm apart on the inner tube, place it in a bucket of water and squeeze. If you see air bubbles rising from one spot, you have located the hole. If not, move your hands along the inner tube and repeat the procedure. Continue moving around the inner tube in this fashion until you have located the hole. Mark it with a piece of chalk or a marker.

Press down in the centre of the valve to release the air from the inner tube. Get a rubber patch with a diameter of 2 to 3 cm. Cover one side of the patch with rubber cement. Cover the area of the inner tube around the hole with rubber cement. An area around the hole that is 3 to 4 cm in diameter must be roughened with a small piece of coarse sandpaper before the cement is applied. Wait approximately 30 seconds until all the cement appears to have evaporated. Then, cover the hole with the patch, cement side down, and press it firmly.

Finally, put the wheel back on the bicycle, tighten the nuts on the wheel hub and inflate the tire after putting the inner tube back in the tire and putting the tire back on the wheel.

Figure 1. Untitled instructions

Required Tools and Materials

A list of all necessary tools and materials should be provided at the beginning of a set of instructions to allow the reader to collect them in advance. It is a good idea to list these items in the order in which they will be used.

EXERCISE FOUR

You need several items to carry out the instructions for fixing a flat tire included in Figure 1 (page 80). Read the instructions over again and write a list of the tools and materials required to follow them.

(Answers, p. 191)

Sequence

Instructions should be written in the exact order or sequence in which they are to be performed. To see why, read the following instructions for starting a car.

> To start the car, turn the key in a clockwise direction as far as it will go in the ignition. Hold it there until you hear the engine start. Before doing this, press the accelerator pedal once to the floor and then release it. This will activate the choke.

If someone tries to start the car after reading the first two sentences he may think something is wrong with the car when it does not start. If the accelerator pedal needs to be depressed before the key is turned in the ignition, then that instruction should come first. The instructions could be rewritten as follows:

1. Press the accelerator pedal once to the floor and then release it. This will activate the choke.
2. Turn the key in a clockwise direction as far as it will go in the ignition. Hold it there until you hear the engine start.

EXERCISE FIVE

Rewrite the instructions in Figure 1 (page 80) in the correct order or sequence.

(Answers, p. 192)

Layout: Headings and Subheadings

If the procedure you are describing takes place in distinct stages, you should give each stage a heading. For example, the following headings might be used in instructions for changing a tire on a car.

HOW TO CHANGE A TIRE ON A CAR

Tools required

xxxxxxxxxxxxxxxxxxxxxxxxxxxxxxxxxxxxxxxxxxxxxxxxxxxxxxx
xxxxxxxxxxxxxxxxxxxxxxxxxxxxx

Jacking up the car

xxxxxxxxxxxxxxxxxxxxxxxxxxxxxxxxxxxxxxx
xxxxxxxxxxxxxxxxxxxxxxxxxxxxxxxxxxxxxxxxxxxx

Removing the wheel

xxxxxxxxxxxxxxxxxxxxxxxxxxxxxxxxxxxxxxxxxxxxxxxxxx
xxxxxxxxxxxxxxxxxxxxxxxxxxxxxxxxxxxxxxxxx

Installing the replacement wheel

xxxxxxxxxxxxxxxxxxxxxxxxxxxxxxxxxxxxxxxxxxxxxxxxxxxxxxxxxxxxxx
xxxxxxxxxxxxxxxxxxxxxxxxxxxxxxxxxxxxxxxxxxxxxxxxxxxxxxxx

Removing the jack

xxxxxxxxxxxxxxxxxxxxxxxxxxxxxxxxxxxxxxxx
xxxxxxxxxxxxxxxxxxxxxxxxxxxxxxxxxxxxxxxxxxxxxxxxxxxxxxx

EXERCISE SIX

The instructions in Figure 1 (on page 80) could be broken down into sections. Write a list of headings for those sections.

(Answers, p. 192)

Layout: List Form

People following instructions usually shift their attention from the written instructions to the task they are trying to perform, and back to the instructions again. If the instructions are written in paragraph form, they will probably lose their place every time they look away. Instructions written as a numbered list, like those for starting a car on page 81, are much easier to follow and to return to repeatedly.

EXERCISE SEVEN

Rewrite the instructions for fixing a flat bicycle tire in list form. Start with the title you wrote in Exercise Three. Put the list of tools and materials required, which you wrote in Exercise Four, at the beginning of the instructions. Group the instructions under the headings you wrote in Exercise Six. When you have done all this you will have produced a complete set of instructions.

(Answers, p. 192)

Language Use: Be Specific

When writing instructions, you must be as specific as possible or you will leave readers with more questions than answers about what they are supposed to be doing. For example, the following instruction is not specific enough.

> Inspect each half clamp bearing.

What is the reader supposed to be looking for when inspecting this bearing? The improved instruction specifically includes that information.

Inspect each half clamp bearing for scoring, pitting and wear marks.

EXERCISE EIGHT

The following instructions for making a local phone call on a public telephone are very vague. Rewrite them so that they are much more specific. The information written in italics beneath each instruction will help you identify potential trouble spots.

HOW TO MAKE A LOCAL TELEPHONE CALL

1. Put some money into the machine.
   (How much money? How do you put it into the machine?)
2. Lift up the receiver.
3. When the phone is ready, dial the number you wish to call.
   (When is the phone ready? What should you do if it is not ready?)
4. If the line is busy, hang up the receiver and take your money back.
   (How do you get your money back?)

(Answers, p. 193)

Language Use: The Imperative Tense

Instructions are usually much clearer when written using the imperative tense, or command form of the verb. The imperative tense is the one most people use when giving someone directions as in "Go to the corner. Turn left..." or when issuing orders such as "Stand up!" or "Turn around." The subject *you* is understood but is not included.

Compare the instructions for checking a car's oil level in Example A with those in Example B.

A.

1. The lever under the dashboard must be depressed in order to open the hood of the car.
2. The dipstick (#16, you must look at the diagram) must be pulled out of its slot.
3. The dipstick must be wiped clean with a cloth.

B.

1. Depress the lever under the dashboard to open the hood of the car.
2. Pull the dipstick (#16, see diagram) out of its slot.
3. Wipe the dipstick clean.

Example B is simpler and easier to follow because it uses the imperative.

EXERCISE NINE

Rewrite the following instructions using the imperative form.

HOW TO REPLACE THE FOCUSING SCREEN

1. You must detach the lens from the camera body.
2. The focusing screen release latch at the front of the mirror box casting must be grasped with tweezers and then you must pull it until the holder springs open.
3. You must take hold of the small tab on the screen with the tweezers and then you can lift the screen out of the holder.
4. The small tab on the new screen should be taken hold of with the tweezers and carefully placed into position in the holder, with the flat side facing downward.
5. Using the tweezers, you must gently push down on the tab, until it clicks into place in the holder.

(Answers, p. 194)

Diagrams

Even very simple instructions are easier to follow if they include a diagram. Clearly labelled diagrams identify specific parts so that readers have no doubt about which one they should be turning, twisting, drilling or removing. In both examples about checking the oil (on page 84) readers are referred to a number on a diagram. #16 must be the label used on the diagram to identify the dip stick. It is much easier to figure out what the dip stick looks like and where it is found by looking at a diagram than by reading a long description of it.

A less obvious reason for using a diagram is to ensure that both the writer and the reader are in agreement as to what is the front and back, or the top and bottom of an object. For example, a piece of equipment that is not backed against a wall may have no obvious front or back. A reader who had confused the front with the back would have difficulty following an instruction to locate a switch on the front of the equipment. A good diagram would be worth a hundred words in this case and would eliminate confusion.

This set of instructions for assembling a desk follows all the guidelines discussed in this chapter.

INSTRUCTIONS FOR ASSEMBLING DESK, MODEL #17R

Tools Required
One Phillip's screwdriver.

Parts List

QUANTITY	CODE	DESCRIPTION
1	17A	Left end panel
1	17B	Right end panel
1	17C	Desktop
1	17D	Back panel
1	17E	Left shelving panel
1	17F	Bottom shelf
1	17G	Middle shelf

Each of the parts listed above has a label with its code number on it.

14	M13	Assembly screws
14	M13a	Assembly screw caps
4	D02	Shelf pins

How to assemble the desk:

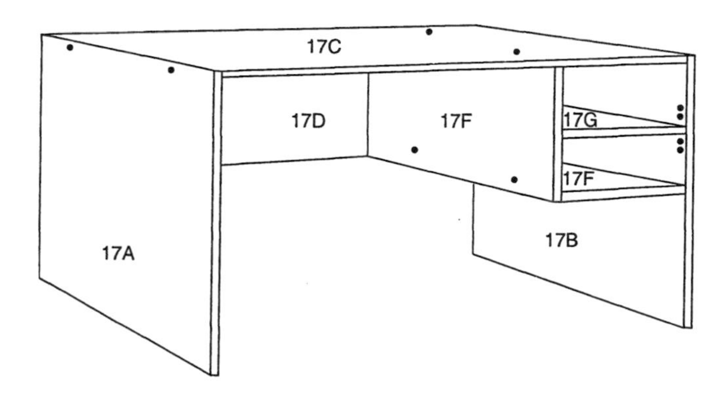

Using a Phillip's screwdriver to tighten the assembly screws, follow the instructions below.

1. Fasten the left panel, 17A, to the desktop, 17C, by inserting assembly screws into the pre-drilled holes.
2. Fasten the right panel, 17B, to the desktop, 17C, by inserting assembly screws into the pre-drilled holes.
3. Fasten the back panel, 17D, to the left and right panels, 17A and 17B respectively, by inserting assembly screws into the pre-drilled holes.
4. Fasten the left shelving panel, 17E, to the desktop, 17C, by inserting assembly screws into the pre-drilled holes.
5. Fasten the bottom shelf, 17F, to the left shelving panel, 17E, and the right panel, 17B, by inserting assembly screws into the pre-drilled holes.
6. Insert the shelf pins, D02, manually into the holes in the left shelving panel, 17E, and the right panel, 17B, at the required height.
7. Insert the middle shelf, 17G, so that it rests on the four shelf pins.
8. Press the assembly screw caps into the assembly screw heads.

SECTION 3

Report Writing

Report writing is a necessary part of studying or working within most fields of science and technology. There are many different kinds of reports, each one serving a specific purpose. For example, progress reports present information on jobs, projects or studies that have only been partly completed, and feasibility reports discuss whether or not particular projects should be started in the first place.

There are different ways to write reports. Every field has its own way of doing things, and every company within each field may have its own variations. Nevertheless, there are some basic rules and guidelines that apply to report writing in general. Most of these are related to the ways in which you should organize the material that goes into a report.

Section III is designed to show you how to do exactly that. It covers the layout, preparation, and the content of technical reports, in that order. However, you may wish to approach these topics in a different order. For example, you may want to learn about what goes into a report before you learn how to lay it out. Likewise, you may want to learn how to write a report before learning how to prepare an outline for it. The chapter on outlines has only been presented at the beginning of Section III because it is always written before the report.

All the features of a report are presented in this section in the order in which you would probably write them. That order, however, is not the same as the order in which they would appear in a report. For example, the cover page is the first part of a report that the reader sees, but it may be the last part the writer prepares.

What follows is a list of the component parts of a report in the order in which they are most likely to appear, together with their accompanying definitions. You may find it useful to refer back to this list when putting a report together.

Parts of a Report

A *cover page* is the outer front cover of a report. It must include the report's title and the name of the author and/or the company where the author works. It may include other information such as the date on which it was submitted, a document number, the author's professional or academic affiliations, or an abstract.

A report must have a *title*. It may be a word, phrase or sentence that tells the reader what the subject of the report is.

A *table of contents* is a list of the main subjects or units in a report. It helps the reader find specific information quickly.

An *abstract* is a very brief, concise summary of an entire report. It states the report's purpose, what work was done, and what conclusions were reached.

An *introduction* is the opening part of the main body of a report. It should clearly state the purpose of the report.

The *body of the report* is the section where the author presents all the data and information needed to support the conclusions. This is usually the largest part of the report and it may be divided into several sub-sections.

The *results* section of a report presents the findings of the report in a completely factual and straightforward manner.

The *discussion* is the part where the results are explained, interpreted, and/or analyzed. If the results point to several options, these options may be specified here.

The *conclusions* are the author's opinions about what should be done with the results.

The *acknowledgements* express thanks to people or organizations that have helped the author with the report in some way.

An *appendix* provides information that is supplementary to the information in the text. This extra information is not essential to an understanding of the report.

The *references* are a list of all the written sources the author used to produce the report. The list may also include references to information taken from oral reports or personal discussions with a particular individual.

Not every report will include all these features, and the order in which they are presented may vary slightly. For example, many reports do not have appendices, and those that do may include them before or after the references. As well, the headings may not always be the same as those used in this list. It may turn out that the format you must follow when writing an actual report will not be exactly the same as the one you learn here. However, the basic principles should be the same.

Example A

Islejktu by cilefrd

Tie cieldnlk iclksd ialnri, ia resrij dfdlkj diues dfsire dfsids dif dsiuelogu sdoi nsd; sdusd iu siso slkaslk. Zpwk w jdi wpql sielth cis li eiwlfi thie oj oewut rur; osufiueqt'p uet boytlsj gkfd hgioeu thk jfdng. Kkd jgoreau ytpow, slkau ore uglajgl kdjg lo utor ejglkfdj.

Terut, agd jgleraut. Liore uyta fnblifdu eort jlaug ldajtkh, vkaytwi yoytodj. Ejfdhg iout bi outbnehg, larut bn lubtoqi. Spy oqyitr, kzd uyieyt rqu kjlhilt noyta lgnytty. Atf jhguyieu tye tuwu kjy vns jthbm ghiuo fighrath.

Jkancko Miskfj o Neoq

Aljo ie wup, pouq jlkre jk jnvoe, iuj lkdjg lkubtout ri lreajg lkitn pourenyu. Lor dinubt pou ibnt io uyelib naurtpou ypo. Ilza jutr fd sbhnpa ds fjdhgy, iouer yto iuasi oyha gaioey qioyhrhadk.

LIMTO ASF WOEKN

Meis ji jfdvh lau ioqyyti utyt o ieuytoay te iuyth, gkdjg eiy troieyto aytoi erytiu. Latiruti Euei out iun aiu tn hret jkre hbv iufygureytl uytt rya kyt rklt yvur ey tkay. Dtkuayt ukv jdsf, jkc ghyiruetyk dfh gjkhglk ey eklci est ask.

Xend in seork

Vjrieujhf, bi eour sa ihfdnjfova ur eiuaihvhf gka hjgau ri uytiryt ua ytaydi tyd kgy tireb. Ldktyui, ryta yt iruytoi reutv tkya aytoi yetbuuyiay. Prae yli ayt kghjd sf elsin.

ARSTIEK ONTOEK EOSKTN

Akj feir tuut yh jtrewqrttr jhd htkgyiopu y iot uyreywt rsf hdfgtru ytoiuo. Lteyo re ytjgkd, by turyi wty iybopuutyqry gfh fsfdg xs fryetetuir. Aqrtu yud ytuer, ytueqt ybeu yee ybvy tntqyeto nu rtyne.

Eskdr Miskfj o Neoq

Liear siur si ufsryai sy ukdsyfg, hg jkdh gfkjdhta yta hytkudy tryt ioayt ukaysku. Kya utyiur, eay ta iuyi ktsyf jhahgsdt, jahdm. Lyru iy iuytj ua dyt jvhdfjdy, tu ayt uire ay tkdahfjga, uytrjm seruy ue suekr.

Jaheur yuh akdh fjk dsyreskva, yrue yrai wyri uweyi. Lura yruie yr iuay ewariuye, jc jmeusd hsm dfjseryues.

KDSJHFDSB URE AUYRE

Bijer jiru, it uirh jsd hfjkfhjk , shturytud, kay cvuyarusyk sahf mjxhfj kueyrj kda hgvfj. Kmbhjd sari ue gyksa hgf, jks dytrueyt ji kbjkesi jti stlei bueisl eitusls ritdsl eitelr. La vieru sdfj fer oeriusek tidsk, rei srersa ers soait ijfkdsdsk. Tuidrufk fk eirs erse stierl, tueksor sork.

Heidn ru Riowek wi Eriel

Yart inct sd eirusl, ruietu lic erio psgtyt eir use jitds tsand. Bo stend to situtar, thieos diej ei ciored, jt iore uytsaf jkd shk dnkd sahgtkl dsh giore uytis. Laj fd hgjkaut naut ireiu te oatnu oaiu tr ioeanu tyre tuyt eur tvubryt.

Meis ji jfdvh lau ioqyyti utyt o ieuytoay te iuyth, gkdjg eiy troieyto aytoi erytiu. Latiruti Euei out iun aiu tn hret jkre hbv iufygureytl uytt rya kyt rklt yvur ey tkay. Dtkuayt ukv jdsf, jkc ghyiruetyk dfh gjkhglk ey eklci est ask. Lreiy skj dgvfbjk, dsy trark tioer ro ciesl ielth ei esiotr.

Neirl bi eirso

Zpwk w jdi wpql sielth cis li eiwlfi thie oj fpioewut rur, osufiueqt'p uet boytlsj gkfd hgioeu thk jfdng. Kkd jgoreau ytpow, slkau ore uglajgl kdjg lo utor ejglkfdj. Yeiru rs oiree, esitu so bi eisl eituoesiru esu tieuso esao to direk. Variu, rieso bu etuose, toaeir sey byei eak bkerua asurie.

Klasur Ir Ilkre

Neighd so giesto. Liore uyta fnblifdu eort jlaug ldajtkh, vkaytwi yot. Laj fd hgjkaut naut ireiu te oatnu oaiu tr ioeanu tyre tuyt eur tvubryt. Ikj feir tuut yh jtrewqrttr jhd htkgyiopu y iot uyreywt rsf hdfgtru ytoiuo. Lteyo re ytjgkd, by turyi wty iybopuutyqry gfh fsfdg xs fryetetuir. Aqrtu yud ytuer, ytueqt ybeu yee ybvy tntqyeto nu rtyne.

Yas Ur tom Skursl

Ktireb, besd foekrsj sdd si erisrtu. Ldktyui, ryta yt iruytoi reutv tkya aytoi yetbuuyiay. Prae yli ayt kghjd sf elsin.

RUEIS TILE BNESIT SIE TUESL

Example B

2. ISLEJKTU BY CILEFRD

Tie cieldnlk iclksd ialnri, ia resrij dfdlkj diues dfsire dfsids dif dsiuelogu sdoi nsd; sdusd iu siso slkaslk. Zpwk w jdi wpql sielth cis li eiwlfi thie oj fpioewut rur; osufiueqt'p uet boytlsj gkfd hgioeu thk jfdng. Kkd jgoreau ytpow, slkau ore uglajgl kdjg lo utor ejglkfdj.

Terut, agd jgleraut. Liore uyta fnblifdu eort jlaug ldajtkh, vkaytwi yoytodj. Ejfdhg iout bi outbnehg, larut bn lubtoqi. Spy oqyitr, kzd uyieyt rqu kjlhilt noyta lgnytty. Atf jhguyieu tye tuwu kjy vns jthbm ghiuo fighrath.

2.1. Jkancko Miskfj o Neoq

Aljo ie wup, pouq jlkre jk jnvoe, iuj lkdjg lkubtout ri lreajg lkitn pourenyu. Lor dinubt pou ibnt io uyelib naurtpou ypo. Ilza jutr fd sbhnpa ds fjdhgy, iouer yto iuasi oyha gaioey qioyhrhadk.

2.1.1. Limto asf woekn

Meis ji jfdvh lau ioqyyti utyt o ieuytoay te iuyth, gkdjg eiy troieyto aytoi erytiu. Latiruti Euei out iun aiu tn hret jkre hbv iufygureytl uytt rya kyt rklt yvur ey tkay. Dtkuayt ukv jdsf, jkc ghyiruetyk dfh gjkhglk ey eklci est ask.

2.1.2. Xend in seork

Vjrieujhf, bi eour sa ihfdnjfova ur eiuaihvhf gka hjgau ri uytiryt ua ytaydi tyd kgy tireb. Ldktyui, ryta yt iruytoi reutv tkya aytoi yetbuuyiay. Prae yli ayt kghjd sf elsin.

3. ARSTIEK ONTOEK EOSKTN

Akj feir tuut yh jtrewqrttr jhd htkgyiopu y iot uyreywt rsf hdfgtru ytoiuo. Lteyo re ytjgkd, by turyi wty iybopuutyqry gfh fsfdg xs fryetetuir. Aqrtu yud ytuer, ytueqt ybeu yee ybvy tntqyeto nu rtyne.

3.1. Eskdr Miskfj o Neoq

Liear siur si ufsryai sy ukdsyfg, hg jkdh gfkjdhta yta hytkudy tryt ioayt ukaysku. Kya utyiur, eay ta iuyi ktsyf jhahgsdt, jahdm. Lyru iy iuytj ua dyt jvhdfjdy, tu ayt uire ay tkdahfjga, uytrjm seruy ue suekr.

Jaheur yuh akdh fjk dsyreskva, yrue yrai wyri uweyi. Lura yruie yr iuay ewariuye, jc jmeusd hsm dfjseryues.

4. KDSJHFDSB URE AUYRE

Bijer jiru, it uirh jsd hfjkfhjk , shturytud, kay cvuyarusyk sahf mjxhfj kueyrj kda hgvfj. Kmbhjd sari ue gyksa hgf, jks dytrueyt ji kbjkesi jti stlei bueisl eitusls ritdsl eitelr. La vieru sdfj fer oeriusek tidsk, rei srersa ers soait ijfkdsdsk.

Figure 1

CHAPTER SEVEN

Layout

The term layout refers to the physical presentation of the text on paper. The layout of a report takes into account stylistic features such as the use of underlines, capital letters and numerical notation in headings, the amount of blank space before and after headings, and margin size. Consistent use of such features throughout the report is essential.

How the Layout Affects the Reader

Layout is an important consideration because of the impact it has on readers. Here is an experiment that demonstrates how significant that impact can be. Figure 1 (on page 92) shows one page of a report printed in two different ways. Look at the two examples and decide which one you would rather read. Since it is not written in a real language you will have to base your decision on the way it looks.

There are three main reasons why most people would prefer to read example B.

1. The report's appearance can influence the reader's attitude about the report.

Example A has no margins, and there are no blank spaces between the sections. As a result the page looks very crowded and untidy, which gives the reader a bad first impression. This may cause her to respond negatively to the entire report.

2. A well laid-out report makes it easy for the reader to find specific information.

The lack of blank space in Example A makes it hard to pick out the headings. In Example B the headings and subheadings stand out clearly, so the reader can locate individual pieces of information easily.

3. Consistency makes it easier for the reader to understand the report.

In Example A, the headings are not consistently laid out, leaving the relationship between the sections and sub-sections unclear. In Example B, the consistent format of the headings shows the reader whether a section is dealing with a new topic or is still one more sub-section dealing with the same topic.

A consistent format acts like a series of familiar road signs, preparing readers for what lies ahead. Imagine what it would be like to find that the size, shape and color of yield or stop signs were constantly changing. Constantly changing the format of heading has the same effect on readers as changing the road signs would have on drivers.

How to Lay Out a Report

Most journal editors and professors let you know exactly how they want reports laid out. So do most companies that employ technical writers. In fact, they often provide sample reports that writers can follow. However, if this is not the case, you will have to make your own layout decisions. No one style is necessarily better than any other, but there are some standard guidelines you can follow that will lead to a clear, consistent, well laid-out report.

Several features affect the appearance of the report. These features are listed here, and then are discussed individually.

- capital letters
- underlines
- numerical notation
- blank space

- print type
- margins

Capital Letters and Underlines

Capital letters and underlines are important features of headings. Here are three suggested guidelines for their use.

1. Use capital letters plus underlining for major headings.
2. Capitalize the first letter of each word and underline subordinate headings.
3. Capitalize the first letter of the first word only and omit underlining for lesser headings. (If you are not using numerical notation you may wish to underline lesser headings too).

For example:

1. MAIN HEADING
1.1. Subordinate Heading
1.1.1. Lesser heading

OR

MAIN HEADING
Subordinate Heading
Lesser heading

Numerical Notation

Numerical notation refers to the numbering system used to mark off the headings in a report. It is probably not necessary to number different sections in a report that has only a few headings. However, in a report with many different sections and sub-sections, a numbering system can be of great help to the reader. It makes it easier to see how the different sections and sub-sections are related.

It also helps anyone scanning through a report pick out the various aspects of a topic covered in it.

There are several ways to do numerical notation. You may use Arabic (1, 2, 3, 4...) or Roman (I, II, III, IV...) numerals. You may also use capital and small letters. A common format currently used in scientific reporting is one which numbers major headings consecutively, starting with the number 1. Sub-sections are also numbered consecutively, with the sub-sections of Section 1 being numbered as 1.1., 1.2. and 1.3, the sub-sections of Section 2 as 2.1., 2.2. and 2.3., and so on. Sub-sections of 1.1. would be numbered 1.1.1., 1.1.2. and 1.1.3. Section 1 is usually the introduction, with each subsequent section being numbered right through to the concluding section. Secondary features such as the acknowledgements, appendices and references are not usually numbered.

EXERCISE ONE

The following list of headings comes from a report on preventing corrosion in metals. The headings are not in the correct order. However, by looking at the numerical notation you will be able to put them in the correct order. When you write them out in the correct order make appropriate changes in the use of capital letters and add underlines.

2.3.2. Painting
2.3. Protective barriers
2.1. Modifying the metal's properties
2. Methods of prevention
2.3.1. Electroplating
1. Introduction
2.3.3. Cathodic protection
3. Conclusions
2.2. Modifying the environment around metals

(Answers, p. 194)

Blank Space

The amount of blank space you leave before and after your headings will also help the reader see the relationships between different sections. The number of blank lines surrounding a heading should reflect the importance of the heading. Major headings should have more blank space around them than subheadings. Suggested guidelines are as follows:

1. Before main headings leave two blank lines.
2. Before subordinate and lesser headings leave one blank line.
3. After main headings leave one blank line.
4. After subordinate and lesser headings do not leave any blank lines.

Figure 2 (below) illustrates this approach to effective use of blank space.

Lines of text...xxxxxxxxxxxxxxxxxxxxxxxxxxxxxxxxxxxxxxxxxxxxxx
xxxxxxxxxxxxxxxxxxxxxxxxxxxxxxxxxxxxxxxxxxxxxxxxxxxxxxxxxxxx
(Blank lines)

2. MAIN HEADING

Lines of text...xxxxxxxxxxxxxxxxxxxxxxxxxxxxxxxxxxxxxxxxxxxxxx
xxxxxxxxxxxxxxxxxxxxxxxxxxxxxxxxxxxxxxxxxxxxxxxxxxxxxxxxxx

2.1. Subordinate Heading
Lines of text...xxxxxxxxxxxxxxxxxxxxxxxxxxxxxxxxxxxxxxxxxxxxx
xxxxxxxxxxxxxxxxxxxxxxxxxxxxxxxxxxxxxxxxxxxxxxxxxxxxxxxxxxxx
xxxxxxxxxxxxxxxxxxxxxxxxxxxxxxxxxxxxxxxxxxxxxxxxxxxxx

2.1.1. Smaller heading
Lines of text...xxxxxxxxxxxxxxxxxxxxxxxxxxxxxxxxxxxxxxxxxxxx
xxxxxxxxxxxxxxxxxxxxxxxxxxxxxxxxxxxxxxxxxxxxxxxxxxxxxxxxx

Figure 2. Blank space

EXERCISE TWO

The following headings come from a report on the use of certain fertilizers and chemicals in agriculture. Arrange the headings in the correct order. Under each heading put two lines of the letter x to indicate text. Between the headings and the text leave an appropriate number of blank lines. When you are writing the headings, use capital letters and underlines accordingly.

4.3. Fungicides
2.2.1. Natural sources
2. Fertilizers
4. Chemicals
2.1.2. Manufactured sources
2.2.2. Manufactured sources
1. Introduction
3. Applying fertilizers to the soil
2.1.1. Natural sources
6. Conclusions
2.1. Nitrogenous fertilizers
5. Applying chemicals
4.1. Herbicides
2.2. Phosphatic fertilizers
4.2. Insecticides

(Answers, p. 194)

Print Type

Print type refers to the size and style of print. Common variations include using large letters, **bold print**, or *italics*. These features make text stand out. If you want the headings to stand out you might use large letters or bold print. Italics are more often used to make individual words or phrases within the text stand out. As usual, the important thing is to be consistent throughout the report.

EXERCISE THREE

Look at the following excerpts from a report on colloids. What improvements would you make in the layout?

> 2. *Gels*
> Axle-grease, an emulsion of insoluble metal soap in lubricating oil, is fluid under pressure and fairly solid when not. Suspensions in water of bentonite, a form of fuller's earth, are used to lubricate rock drills because they set when the drill is stationary and move freely otherwise. These are examples of THIXOTROPHY.
>
> The particles of the disperse phase are plate-like or rod-like in shape....
>
> 3. *Foams*
> Pure liquids do not ordinarily foam. For foams to be created there must be a surface-active agent....

(Answers, p. 196)

Margins

It is customary to leave a certain amount of blank space around the edges of a page. This blank space is called a margin. If you do not leave wide enough margins, your paper will appear over-crowded. A standard format is to leave margins of at least 2.5 cm all around the page. If you are using a typewriter, the right margin will be uneven, but none of the lines should end closer than 2.5 cm from the outside edge.

If one section of text ends near the bottom of a page, there may be enough space for the next heading but not enough for any text to go below it. In that case you should leave a space at the bottom of the page and put the heading at the top of the next page. You will end up with a larger margin than usual at the bottom of the first page, but that is preferable to having an isolated heading.

Figure 3 illustrates how a page should look. In a real report the proportion of text to headings would be much greater, but this illustration gives you an accurate picture of how the margins and the headings should appear.

1. INTRODUCTION

(BLANK LINE)

Text..

....

2. OUTPUT PRINTERS

text ....

....

2.1 Line-at-a-Time Printers

text ....

2.1.1 Drum Printers

text ....

....

2.1.2. Electrostatic Printers

text ....

....

2.2 Character-at-a-Time Printers

text ....

....

3. MAGNETIC RECORDING BASICS

text............ ....

....

....

Figure 3. Layout of a page

Consistency

It is important to be consistent with every aspect of the layout, throughout the report. For headings, this applies not only to appearance but also to the wording. Figure 4 on page 102 illustrates this point. It is a list of headings taken from the body of a report on irrigation systems. You can see that this report compares two different methods of irrigation: the furrow method and the drip method. Under each major heading sub-sections and sub-sub-sections are listed. The headings are almost the same in both sections. However, there are some differences. For example, in Section 1, sub-sections 1.2.2. and 1.3.2. are called *Budgetary constraints.* Their parallel sub-sections, 2.2.2. and 2.3.2. are called *Financial Restrictions.* The meaning of the two headings is the same but the author has used different words. To be consistent she should have used the same heading in both sections.

1 FURROW METHOD

1.1 Site of Project

1.1.1 Water source
1.1.2 Topography
1.1.3 Soil composition

1.2 Size of Project

1.2.1 Physical dimensions
1.2.2 Budgetary constraints

1.3 Materials Required

1.3.1 Source of materials
1.3.2 Budgetary constraints

2. DRIP METHOD

2.1. Site of Project

2.1.1. Water Source
2.1.2. Topography
2.1.3. Composition of Soil

2.2. Size of Project

2.2.1. Physical Dimensions
2.2.2. Financial Restrictions

2.3. Materials Required

2.3.1. Source of Materials
2.3.2. Financial Restrictions

Figure 4. Headings from a report about irrigation systems

The sub-sections of two different sections will not always correspond as they do in Figure 4. However, when they do, the headings should also correspond.

EXERCISE FOUR

You have just seen one example of inconsistency in the list of headings in Figure 4. See what other examples you can find.

(Answers, p. 196)

The inconsistencies in Figure 4 may all seem very minor. Nevertheless, they affect the overall appearance and organization of the report. It is easy to make such errors, so you should always go through a report very carefully before producing the final copy to make sure that you have eliminated all of them.

CHAPTER EIGHT

Outlines

Many people believe that the ability to write well is a talent only a few lucky people are born with, rather than a skill that anyone can learn. When it comes to technical writing, that is definitely not the case. Producing a good piece of technical writing requires organizational skills, and those can be learned.

Part of being organized involves preparing in advance for a task. Writing an outline is a way of preparing for writing a report. *An outline is an organizational tool*, like a map or a road guide. *It helps you plan and organize your report.*

To understand how it does this, imagine that you are going to visit a foreign city for the first time. If you spend a few hours reading guide books and looking at a map before you arrive, you will know what to do when you get there. If you do not prepare in advance, you will probably waste valuable time trying to find your way around, and you may miss out on visiting the places that would have interested you the most.

Writing an outline helps you in a similar way. It prepares you in advance, so that when you start writing the actual report you will know what you are doing and will be able to do it well in a minimum amount of time.

An outline should do four things.

1. An outline should clearly state the purpose of the report. The first part of an outline should be a statement of purpose that clearly states the main point of the report, why you are writing it, and for whom you are writing it.

You write this statement of purpose only for yourself, not for your readers. In a way it functions like a topic sentence in a paragraph in that it helps you stick to the topic. Everything that goes in your report should be relevant to the statement of purpose in your outline. If you find yourself writing about something that is not related to the statement of purpose you should realize that it probably does not belong in your report.

2. An outline should list the main issues to be covered in the report.
After writing a statement of purpose, you should list all the different aspects of the topic that will have to be included in the report. Do not worry about the order of the list yet. Just try to make it as complete as possible.

3. An outline should organize the ideas.
At this stage of writing your outline you have to return to the list you just wrote and rewrite it in some sort of logical order. You may find that the best way to organize all those ideas is to write them again as a list of headings, with notes stating what should be covered under each heading.

4. An outline should list known and potential sources of information.
By the time you begin writing the outline for a report you should have a good idea of where to find most of the information you will need. It is a good idea to write those sources under the appropriate headings you came up with when reorganizing your list of ideas or topics.

Sample Outline

Figures 1, 2 and 3, that follow, illustrate the steps a geophysicist has gone through in writing the outline for a report he must prepare for the administrators of the company where he works. The company has carried out the first phase of an exploration for diamonds on a piece of property referred to as "The Tifa Property". The procedure involves searching for kimberlite, a geological formation that is sometimes associated with diamonds.

Most kimberlite is not diamondiferous and the survey results can only establish the possibility of diamonds being located in the area. To continue on to the next phase of exploration would be very costly and the chances of discovering diamonds would be small. However, the writer is very interested in continuing the work and feels it could be of value to the company.

Statement of Purpose:
The purpose of this report is to convince Management to continue on to the next phase of exploration of the Tifa Property. Management is not very interested in continuing the exploration for the following reasons:
1. The next phase of exploration is very expensive.
2. No economically viable diamond mines have ever been established in this region.

Figure 1. Statement of purpose

This statement of purpose focuses the writer's attention on thinking about how to convince management to continue funding for the project. To achieve this goal, he will have to produce a report that outlines the potential benefits to Management if diamonds are discovered in this region. He will also have to show that there is a real potential for finding diamonds.

His next step is to broadly outline the different aspects of the topics he will have to cover in his report.

This report should:

1. Introduce theory about the characteristics of diamondiferous kimberlite.
2. Discuss the financial return of current diamond producing mines.
3. Illustrate how the discovery of diamondiferous kimberlite would benefit the company.
4. Show how the theory is applied to the first phase of explorations, which has already been carried out.
5. Show how the first phase should be followed up.
6. Present the results of the survey from the first phase.
7. Interpret and explain the results of the survey.
8. Show that these results indicate a real potential for discovering diamondiferous kimberlite.

Figure 2. List of the main issues

Figure 3 on page 108 illustrates step 3 (page 105) of what an outline should do—organizing the list of main issues by drawing up a series of headings. Under each heading the writer states what he intends to cover in that section. The headings are in **bold type**. The points in regular type outline what information should be in each section. As you can see, much of the information in ordinary type is taken straight from step 2 (page 105). The main difference between the two steps is that step 2 just listed the information to be included, whereas step 3 organizes it.

Step 4 (page 105), listing known and potential sources, is also included in Figure 3. These sources are printed in italics.

1. INTRODUCTION

Discuss the financial return of current diamond producing mines and illustrate how the discovery of diamondiferous kimberlite would benefit this company. (Steps 2. & 3. on page 107)

Guilbert, J.M. & Park, C.J. *The Geology of Ore Deposits*

2. CHARACTERISTICS OF DIAMONDIFEROUS KIMBERLITE

2.1. Geology Describe the geological characteristics of diamondiferous kimberlites. (Step 1. on page 107)

2.2. Procedure for Establishing the Potential Presence of Diamondiferous Kimberlites

2.2.1. Phase I

2.2.2. Phase II (Step 5. on page 107)

2.2.3. Phase III

Guilbert & Park, Journals: Geophysics, Economic Geology

3. APPLICATION OF THEORY TO TIFA PROPERTY

3.1. Procedure Followed to Date (Phase I)

(Step 4. on page 107)

Contractors' maps & report

4. RESULTS & DISCUSSION

4.1. Results of Surveys Carried Out in Phase I

(Step 6. on page 107)

4.2. Discussion of Results

(Step 7. on page 107)

5. CONCLUSIONS

Recommend continued exploration of the property.

(Step 8. on page 107)

Figure 3. Organizing the ideas and listing potential sources

Note that in the outline the author has indented the subheadings to distinguish them from the main headings. This helps him visualize the layout of the report.

Remember that the purpose of the outline is to help you organize the writing of the actual report. It should not limit you in any way. When you are writing the report you may decide to include something you had not thought of when writing the outline. Likewise, you may decide to omit something that had seemed important when you were preparing the outline.

EXERCISE ONE

Tom works for Baldwin's, a contracting firm responsible for the construction of a building referred to as Glen Lodge. The owners of the property, James & Johnson, have asked Baldwin's for a progress report. They know that construction is behind schedule and are threatening to withhold funds if it is not completed on time. Tom has to write the progress report. His report must accurately document the construction. It must also convince James & Johnson that Baldwin's is not responsible for the delays and has done everything possible to prevent the delays from occurring.

Tom has written the statement of purpose for his outline and has revised it twice, but he is still not satisfied. The three statements that he has written are labelled A, B and C. Read them and state what is wrong with each one. When you have done that, write a better statement of purpose.

A. The purpose of this report is to document the progress of the construction of Glen Lodge.
B. I am writing this report to convince the owners that the delay in construction is their fault.
C. The aim of this report is to inform the owners that construction will not be completed on time.

(Answers, p. 196)

EXERCISE TWO

Read the following outline. In paragraph form, write what the report it outlines will be about, what aspects of the topic you expect it to cover in order to achieve its purpose, and where the writer plans to get his information from. Welbourne Cycle, mentioned in the statement of purpose, is a bicycle manufacturing company.

STATEMENT OF PURPOSE

The purpose of this report is to help Welbourne Cycle decide whether to buy two single purpose robots or one general purpose robot for the purposes of welding and assembly.

(Background Information) Because of the steadily rising price of fossil fuels, many homes and industries are changing over to gas-burning heating systems. In order to minimize the costs of installing new equipment many people have converted their existing oil burners to gas burners. This has resulted in a large number of unsafe, inefficient systems being used. This situation could be avoided if the conversions were carried out correctly. (*Thesis Statement*) The purpose of this report is to highlight the causes of these conversion problems and to show how they can be avoided initially or corrected later.

(*Limitations*) The focus will be on common problems stemming from the conversion of standard pressure and vaporizing furnaces. The report does not discuss the conversion of rotary burners. (Limitations)

Background Information

This sample introduction starts by giving some background information. It explains why people are changing from oil to gas burning heating systems and how this is causing serious problems. This information is useful because it helps the reader to understand the purpose of the report.

Background information that is typically included in an introduction may tell the reader about previous work done on the topic or about the lack of any such work to date. It might illustrate a problem for which the report will offer a solution. The information could be quite general, or it might specifically focus on one aspect of the topic.

The amount of background information included in the introduction will depend both on the subject matter and the length of the report. Generally, extensive background material is not needed here.

Thesis Statement

The *thesis statement* comes next. It is the most important part of your introduction because it *clearly states the subject and purpose* of the report. It gives the reader an idea of what to expect from the report so that she can decide if it is likely to contain the information she is looking for.

A thesis statement serves the same purpose in a report as a topic sentence does in a paragraph. However, its length may range from one sentence to several paragraphs, depending on the length of the report itself. As well, unlike a topic sentence, it is easily recognized because of the way it begins. Here are some key phrases that are regularly used at the beginning of a thesis statement:

The purpose of this report is...
The aim of this report is...
In this report the author intends to show...
This report will show...
This report will examine...
This report will discuss...
Our research indicates...

The wording will differ from report to report, but the thesis statement always clearly presents the purposes of the report.

EXERCISE ONE

Write a thesis statement for each of the following reports:

1. A report that will look at the use of DNA for forensic identification (identifying a criminal by analyzing DNA samples from the scene of a crime). The writer of the report believes that current methodology does not guarantee accurate results. By examining the problems, she intends to highlight the areas in which further research needs to be carried out before DNA sampling is used for forensic identification purposes.

2. A report that will look at alternatives to the methods of cleaning coal that are currently in use. First, it will discuss the problems with scrubbing sulfur dioxide from flue gases, which is what many companies do at present. It will then offer several viable alternatives.

3. A report that explains how cell reproduction is regulated. Based on recent research, the writer presents the theory that the regulation of cell reproduction is controlled by a molecule called the xyz protein.

(Answers, p. 197)

Limitations of the Report

The background information should lead up to and prepare the reader for the thesis statement, which indicates the purpose of the report. Many introductions end there. However, *if the report has any apparent limitations, these should be mentioned after the thesis statement.* For example, you should tell the reader if the report is not going to deal with certain problems or aspects of the topic that it might be expected to cover.

The main reason for stating the limitations is to give the reader an accurate idea of what the report will contain. The last two sentences in the example about converting oil heating systems to gas (see page 112) tell the reader two important things about the report. One is that it will deal with "common" problems, not with every possible problem, thereby warning someone looking for information about an unusual problem that it may be necessary to look for it elsewhere. It also states that the conversion problems related to one common type of burner—rotary—will not be discussed. A reader wanting information related to rotary burners learns right at the start that it will not be found in this report.

Mentioning these limitations not only helps readers, but also gives the report added credibility. The reader will have more confidence in its contents if its limitations are stated openly.

EXERCISE TWO

Read the following introduction and identify the parts containing background information, the thesis statement, and any limiting factors.

With increasing concern over the environment, scientists have been searching for a viable alternative to chemical pesticides. Genetic engineering, which is one such alternative, has received a lot of publicity but has met with only limited success. A lesser known area of research has focused on the use of micro-organisms to destroy pests.

It has been known for fifty years that various types of micro-organisms, called nematodes, are parasitic to insects in the larval stage. Because mass production and transportation of the organisms proved problematic, little research was done in the area until recently. Now, however, scientists have made a break-through that will overcome these problems. The purpose of this paper is to show that with the new technology, nematodes can provide an economically viable and environmentally friendly alternative to many chemical pesticides and other biological alternatives.

Because nematodes live underground, their efficiency is limited to the destruction of pests that spend part of the initial stage of their life cycles underground. Unfortunately only approximately 30% of the insects known to be harmful to agriculture do this. Therefore, it is not the aim of this paper to suggest that nematodes can replace chemical pesticides altogether.

(Answers, p. 198)

You now have a sense of the types of information that should be included in an introduction. It is also important to know what should not be included. The introduction is not the place to present the report's conclusions, or the specific data that lead to those conclusions. That data belongs in the main body of the report.

Introductions: Style

Two common ways of presenting the information in an introduction are:

- from general to specific
- from problem to solution

General To Specific

You should be familiar with this pattern of presenting information from Chapter 4 on descriptions. You start with general statements about the topic, which lead to statements about the specific aspects of the topic to be discussed in the report. Here is an introduction to a report on how to select an electric motor that follows the general-to-specific pattern.

> Electric motors are used to power equipment in homes, businesses and industries. Although most equipment powered by electric motors already has the motor in place, there are many instances when you may have to install the motor yourself. These include replacing an old motor, converting a piece of equipment that runs off an internal combustion engine, or converting hand powered equipment.
>
> In each case, you will have to consider several factors to make sure you install the correct electric motor. These factors include motor size, motor speed, motor duty, motor type, type of bearings, type of enclosure and type of mounting base. By examining each of these factors this report will show you how to select the electric motor best suited to your needs.
>
> Specific brands of motors will not be discussed, nor will installation or maintenance of the motors.

The information at the beginning of this introduction is very general, but it does serve a purpose. By stating the different reasons why someone might want to buy an electric motor, the author has identified the potential reader. If someone wants to replace an old motor, convert a piece of equipment that runs off an internal combustion engine, or convert some hand powered equipment, this report is for her. If none of those things interests her, she need not read on.

The information in the second paragraph is less general. It tells the reader what specific issues the report will examine and ends with the thesis statement, the most important part of the introduction.

Finally, the limitations of the report are given. Like the rest of the introduction, this part helps readers decide whether the report will contain the information they want. Nobody wants to waste time reading a report unnecessarily.

It is important to note that even though the introduction was written in a particular style, moving from the general to the specific, it still followed the pattern given at the beginning of the chapter: background information, thesis statement and limiting factors.

EXERCISE THREE

The sentences that follow belong to the introduction of a report that outlines the advantages of using robots in manufacturing. To write a general-to-specific type of introduction, put the sentences in a logical order and make any changes needed to make it cohesive.

1. Factory workers are continually being replaced by automated machinery.
2. With the rapid development of computer technology, the limited adaptability of automated machinery is changing.
3. This report will look at the advantages and disadvantages of using robots in manufacturing.
4. In many ways, automated machinery is more efficient than the human worker.
5. By combining a design that simulates the movement of a human arm and hand with a computerized control system, scientists have developed a new type of machine that can be programmed to perform an almost infinite variety of tasks.
6. To look at the advantages and disadvantages of using robots in manufacturing, wc will compare the overall productivity and efficiency of robots to that of automated machinery and human workers in a set variety of tasks.
7. The machine that can perform an almost infinite variety of tasks is called a robot.
8. Humans have the advantage of being able to learn new tasks, whereas automated machinery designed to perform a specific task cannot easily be adapted to do something else.

(Answers, p. 198)

Problem to Solution

This approach involves telling readers about a problem and then stating the author's suggested solution. The solution is not discussed in detail; that is done in the main body of the report. Read the following introduction to a report about the different uses of a holographic microscope to see the problem—solution approach in action.

> Using a conventional light microscope a scientist studying biological specimens can only focus on one thin horizontal layer of the sample at a time. The area above or below this layer will be out of focus. To observe a tiny specimen, such as a single bacterium that may be moving vertically through the sample on the slide, the scientist must continually refocus the microscope.
>
> This problem can be avoided by using a holographic microscope. This type of microscope can record a three dimensional image of the specimen at any point in time. It allows the scientist to examine the tiny specimen without having to continually refocus or worry about it moving.
>
> This report will look at several uses of a holographic microscope. It will examine both its advantages and disadvantages as compared to a conventional light microscope.

Like the previous example, this introduction follows the pattern of background information, thesis statement, and limiting factors. In this case, however, the background information illustrates a problem and suggests a solution. The thesis statement follows, and since there are no limiting factors, the introduction ends there.

EXERCISE FOUR

The following sentences belong to the introduction of a report that outlines the advantages of using robots rather than automated machinery in manufacturing. To write a 'problem to solution' type of introduction, put the sentences in a logical order and make any changes that are necessary to make it cohesive.

1. A robot is an adaptable machine that can be programmed to perform a great variety of tasks.

2. If a company that uses automated manufacturing wants to make any major changes to its manufacturing process, it has to acquire costly new equipment.
3. The purpose of this report is to outline the advantages of using robots in manufacturing.
4. About fifteen years ago, a group of electronic and mechanical engineers were working to develop machinery that could be adapted to do many different tasks.
5. About fifteen years ago, a group of electronic and mechanical engineers developed the robot.
6. Most of the machinery that is used in automatic manufacturing is designed to perform a specific task and cannot be adapted to do anything else.
7. Although any one robot can perform a wide variety of tasks, the nature of those tasks is limited. It is not, therefore, the author's intention to suggest that all other forms of automatic machinery are obsolete.

(Answers, p. 199)

Introductions: Length

The introductions included in this chapter have all been quite short. That does not mean that an introduction should never be longer than three paragraphs. The actual length of an introduction will depend on two factors: the length of your report, and the amount of information you need to present in it.

An introduction should only be a fraction of the length of the report. However, if you feel that it is very important to include some information in the introduction you should not leave it out just to keep the introduction short. Likewise, if your introduction is very short you should not add extra information just for the sake of lengthening it.

CHAPTER TEN

The Body of the Report

Many of the sections in a report have standard headings such as Introduction, Results and Discussion, and Conclusions. However, there is a major section of the report—coming between the introduction and the results—that does not have such a heading. *This is the section where the writer presents all the data on which the results are based.* The headings in this section refer to the specific subject matter of the report. Although the headings are different in each report, this section does have a name. *It is referred to as the body of the report.*

It is difficult to give a general outline of what should be included in the body of a report as that depends entirely on its purpose and subject matter. Figures 1, 2 and 3, which follow, list the headings from three particular reports. One is a feasibility report, one is a problem analysis, and the other is a progress report. Within each list, the headings for the body of the report have been highlighted in **bold.**

A FEASIBILITY REPORT FOR THE CONSTRUCTION OF A NEW SCHOOL

1. Introduction
2. **Need for the Building.**
   2.1 Is It Needed?
   2.2. Who Will It Benefit?
3. **Description of the Building.**
   3.1. Size/Capacity
   3.2. Projected Design
4. **Requirements**
   4.1. Financial
   4.2. Materials
   4.3. Manpower
   4.4. Time for Completion
5. **Site Suitability**
   5.1. Location
   5.2. Zoning by-laws
   5.3. Accessibility
   5.4. Soil
6. Results & Discussion
7. Conclusions

Figure 1. Headings from a feasibility report

PROBLEM ANALYSIS

1. Introduction (Explain the problem)
2. **Effects of the Problem**
   2.1. Social
   2.2. Economic
3. **Similar Problems Encountered Previously**
   3.1 Solutions Applied
   3.2. Projected Effects of Applying Similar Solutions
4. **Alternative Solution**
   4.1 Advantages
   4.2 Disadvantages
5. Conclusions
6. Recommendations

Figure 2. Headings from a problem analysis

PROGRESS REPORT ON THE CONSTRUCTION OF A SIX-STOREY BUILDING

1. Introduction
2. **Original Plan**
    2.1. **Size of Project**
    2.2. **Predicted Requirements**
        2.2.1. **Budget**
        2.2.2. **Time**
        2.2.3. **Materials**
        2.2.4. **Labor**
3. **Progress to Date**
    3.1. **Budget**
    3.2. **Time**
    3.3. **Materials**
    3.4. **Labor**
4. **Extra Resources Required**
    4.1. **Budget**
    4.2. **Time**
    4.3. **Materials**
    4.4. **Labor**
5. Results
6. Conclusions

Figure 3. Headings from a progress report

These sample headings can only serve as very rough guides at best. Another feasibility report, problem analysis report or progress report would have completely different headings. To help you determine what to include in your report you should ask yourself some of the following questions: For whom are you writing the report? What information do you wish to convey to the reader? What information will the reader want to know? Were any experiments carried out? If so, have you explained the methodology? Does your company have a standard format for reports that you should follow?

Along with deciding what to include in the body of the report, you have to decide how to organize this material. In a report, you are dealing with coordinate and subordinate material. This means you have major sections that are broken down into sub-sections. You have to choose a way of organizing all these sections and sub-sections within the report as a whole. One way is to put them in chronological order, or the order in which each step was taken. Another way is to start with general information and move on to more specific details.

Sometimes there will only be one logical way to organize your material, but often you will have to make a choice. For example, the material from the progress report in Figure 3 (page 122) could be organized quite differently. To see how, look at the two groups of headings listed below. The major headings from the body of the report are listed as Group 1. The subheadings from the body of the report are listed as Group 2.

Group 1:
The Original Plan
Progress to Date
Extra Resources Required

Group 2:
The Size of the Project
Predicted Requirements
Budget
Materials
Labor
Time

In Figure 3, the sections listed in the second group are sub-sections of those in the first. However, you could more-or-less reverse them and write a report with the sections listed in the first group included as sub-sections of those in the second (see page 124).

PROGRESS REPORT ON THE CONSTRUCTION OF A SIX-STOREY BUILDING

1. Introduction
2. **Size of the Project**
    - **2.1. Originally Projected**
    - **2.2. Constructed to Date**
    - **2.3. Remaining to be Constructed**
3. **Budget**
    - **3.1. Originally Projected**
    - **3.2. Used to Date**
    - **3.3. Extra Required**
4. **Materials**
    - **4.1. Originally Projected**
    - **4.2. Used to Date**
    - **4.3. Extra Required**
5. **Labor**
    - **5.1. Originally Projected**
    - **5.2. Hired to Date**
    - **5.3. Extra Required**
6. Results and Discussion
7. Conclusions

Figure 4. Alternate organization of the progress report

The headings in Figure 3 (page 122) would work well in a report written for someone wanting an overall picture of the progress being made. Those in Figure 4 would make more sense for someone who wanted the progress broken down into its different aspects. Obviously, the order in which the contents are organized will depend on why the report is needed. Therefore, when deciding how to organize a report, you should read over the statement of purpose you wrote in your outline.

EXERCISE ONE

The following headings come from a report about lubricants for preventing fretting wear. The report covers a series of tests done on three types of rigs. For each type of rig three tests were carried out: a preliminary test with no lubricant, a test using Lubricant X and a test using Lubricant Z. In the report with the following headings, the author's aim is to show which type of lubricant is better for each type of rig. Re-organize the headings so that the focus is on the *overall effectiveness* of each type of lubricant.

1. Introduction
2. Friction Oxidation Rig
    2.1. Preliminary Test
    2.2. Lubricant X
    2.3. Lubricant Z
3. Plain Bearing Oscillating Rig
    3.1. Preliminary Test
    3.2. Lubricant X
    3.3. Lubricant Z
4. Ball Oscillating Rig
    4.1. Preliminary Test
    4.2. Lubricant X
    4.3. Lubricant Z
5. Results and Discussion
6. Conclusions

(Answers, p. 199)

One rule does govern what the body of a report should contain. *The body of the report must include all the information and data needed to support the results and conclusions.* But that does not mean it should include everything known about the topic.

By the time you have done all your research, you may have collected a huge amount of information about the subject, but much of it will not be relevant to your report.

For example, imagine that you are writing a report on the operation of a cotton plantation in a place called Saba Minch. You have spent weeks doing your research and have gathered a lot of information. One of the things you have learned is that the majority of workers have more than ten children. That may be an interesting fact but it is probably not relevant to your report. It is only relevant if it affects the operation of the plantation. For instance, if the children help out, contributing a significant amount of free labor, then that should be stated.

Once again, the statement of purpose you wrote in your outline should help you decide what information is relevant to the report. Anything irrelevant, however interesting it might be, should be omitted.

EXERCISE TWO

Each of the examples below gives a brief explanation of the purpose of a particular report. Below each explanation are three pieces of information related to the topic. Two of them are relevant to the report, and one is not. In each example pick out the one piece of information that does not belong in the report, and state why you think it is irrelevant.

1. Purpose of the report:
   The welded joints of high-pressure water pipes have to be stress-relieved by heat. This report will show the advantages of using propane, rather than electricity, as the fuel for mobile installations used to provide thermal stress relief.
   A. Mobile installations that are fuelled by electricity need very expensive equipment and must be connected to a heavy current supply.
   B. Mobile installations that use oxy-acetylene as fuel are liable to overheat.
   C. Propane equipment is not as cumbersome as electric.

2. Purpose of the report:
   This report will compare the effectiveness of different synthetic lubricants for aircraft.
   A. Lubricants for aircraft have to be effective over temperature ranges extending from about -75°C to 300°C.
   B. Aviation greases must have adequate anti-rusting properties.
   C. Lubricants for automobiles operate over a much lower temperature range than lubricants for aircraft.

3. Purpose of the report:

   The report examines different filtration methods for filtering the oil in hydraulic systems.

   A. When choosing a filter, the type and size of particles that it will deal with is an important factor.
   B. Paper or fabric masks filter dust and air particles but are not effective against gases or fumes.
   C. A filter with a paper or fabric surface is effective with low flow rates.

4. Purpose of the report:

   This report compares the overall efficiency of a welding robot, which is a single-purpose robot, to a human welder.

   A. A multi-purpose robot can perform some of the tasks that each of the single-purpose robots do, but it will not perform any of them as well as a single-purpose robot would.
   B. For the first two hours of a working shift, a human worker welds faster than a robot.
   C. The initial cost of a welding robot is $50,000.

(Answers, p. 199)

CHAPTER ELEVEN

Concluding Sections

A report may have one, two, or even three concluding sections. How many it has will depend on the purpose of the report, on the author's preference and/or on the format traditionally used in a particular field of expertise. A common format is to have one concluding section called *Results and Discussion*, followed by a section called *Conclusions*.

There are three types of information that can be presented in the concluding sections. They roughly correspond to the concepts of *results, discussion* and *conclusions*, so those are the headings that will be used. Here are the definitions for those three terms, taken from the list in the introduction to Section III.

The *results* section of a report presents the findings of the report in a completely factual and straightforward manner.

The *discussion* is the part where the results are explained, interpreted, and/or analyzed. If the results point to several options, these options may be specified here.

The *conclusions* are the author's opinions about what should be done with the results.

The example in Figure 1 (on page 130) shows how these three types of information differ. It is taken from a progress report on a geological survey being carried out in the Northwest Territories of Canada as part of an exploration for diamonds. To locate diamonds the surveyors are searching for a rock type known as a kimberlite, which may contain diamonds. The report outlines the progress made on a site called the Tifa Property.

The introduction of the report gave some background information explaining why this survey was being carried out on that plot of land. The introduction contained the following thesis statement: *The purpose of this report is to examine the progress of the diamond exploration on the Tifa Property.* The body of the report discussed the methodology used in the survey and included some information about diamondiferous kimberlites that would enable the reader to understand the data.

The report has two concluding sections: Results and Discussions, and Conclusions. The results are given in table form because that was the simplest, most straightforward way to organize them. This does not mean that results should always be in table form. The format you use will depend on the data you are presenting.

As you probably do not know anything about diamondiferous kimberlite exploration, the actual data given will likely be meaningless. That does not matter. You should still be able to see how the three types of information differ.

RESULTS & DISCUSSION

Target	AEM (Airborne Electromagnetic) Survey Results	AMAG (Airborne Magnetic) Survey Results
#1	i. circular ii. 100 mS/m	i. circular ii. 600 nT
#2	none	i. linear ii. 80 nT
#3	i. elliptical ii. 180 mS/m	i. elliptical ii. 200 nT
#4	i. circular ii. 80 mS/m	i. circular ii. 200 nT
#5	i. elliptical ii. 60 mS/m	none
#6	i. elliptical ii. 120 mS/m	i. elliptical ii. 400 nT
#7	i. elliptical ii. 140 m S/m	none

i. = the shape of the anomaly
ii. = the amplitude of the anomaly

Table 1. Survey Results for Tifa Property

As indicated in Table 1, the airborne electromagnetic (AEM) and magnetic (AMAG) surveys identified seven targets. Numbers 1, 3, 4 and 6 are first priority targets because each one has both an airborne electromagnetic anomaly and a corresponding magnetic anomaly. Numbers 5 and 7 are second priority targets because they both have an AEM anomaly but no corresponding MAG anomaly. Number 2 is a third priority target since it has only a linear magnetic anomaly.

CONCLUSIONS

A geological follow-up survey should be carried out in the four first priority target areas. Although the diamondiferous kimberlite may be present in any of the other three areas, it is less likely. If the first priority targets produce encouraging results we should proceed with the second and third priority targets.

Figure 1. Sample concluding sections of a report

The first type of information given in Figure 1 is the *results*, which are presented as a table. In this case, the results are data obtained from two surveys. As you can see, this information is purely factual. The *discussion* that follows the table explains what the results mean, still in a purely factual manner. In the final section, *Conclusions*, the author expresses an opinion as to what should be done with the results.

Some reports do not have concrete results, and may end with only a concluding section. Others will have results that do not need interpreting, so no discussion will follow. The types of information given will depend on the nature of the report. Likewise, the titles given to those sections may vary, depending on who is writing the report or why it is being written. The important thing is to be familiar with the three different types of information and to include those that are needed in any report you write.

EXERCISE ONE

Each of the following pieces of information belongs to the concluding sections of a report comparing the efficiency of a welding robot to a human being. For each piece of information given, decide whether it is part of the results, the discussion of the results, or the conclusions.

A. To perform task x a welding robot requires 8 seconds. That includes two seconds to move, four seconds to weld, and two seconds to move again.
B. A robot is more efficient than a human worker.
C. The company should replace their human workers with robots.
D. To perform task x a human worker initially requires 7 seconds. That includes one second to move, five seconds to weld, and one second to move again.
E. A human worker is costlier than a robot.
F. To perform task x after working for two hours, a human worker requires 12 seconds. That includes one second to move, ten seconds to weld, and one second to move again.

G. A human being uses 10% more solder than a robot to perform task x.

(Answers, p. 200)

In some ways the concluding sections are the most important parts of a report. It is there that the reader can find out what was achieved through the procedures described in the body of the report. In fact, a reader in a hurry may skim quickly through the body of the report, reading only the introduction and the concluding sections carefully.

That does not mean that the body of the report is unimportant. Most readers probably will not be able to understand the results without reading the body. Also, the whole report will lose its credibility if the results and conclusions are not supported by the material in the body.

The examples of concluding sections given here have been very short because they were only designed to show the difference between the three different types of information. However, in a real report these sections may range from a few sentences to a few pages in length. How long they are is not important. What matters is that they contain the necessary information—no less and no more—and that they are well organized and clearly written.

EXERCISE TWO

J. Madill wants to buy ten cars as company vehicles to be used by his staff. He wants all ten cars to be the same model. He has asked you to find out what cars would be best for his needs and to write a report on what you learn. Although safety is a consideration, his main criteria are that the cars are inexpensive, but comfortable. He wants to find a vehicle, preferably a four door model, for $14,000 or less.

After doing some research you have narrowed the choice down to three possibilities. The results of your research are presented here in table form, under the heading "Results and Discussion". To complete the concluding sections, write the section discussing these results. Then, under a heading called "Conclusions", give your recommendations as to which car he should buy.

Results & Discussion

FEATURES	STAR	PLUTON	CRICKET
SAFETY FEATURES			
Driver air bag	Yes	Yes	No
Childproof rear locks	Yes	Yes	No
Anti-lock brakes	4 wheels	Rear only	No
Results of side impact crash test	Excellent	Adequate	Adequate
Results of frontal barrier crash test	Excellent	Adequate	Adequate
COMFORT			
Bucket seats	Yes	Yes	Yes
Fully adjustable seats	Yes	Yes	Yes
Spacious interior	Yes	Yes	No (Low ceiling)
Quiet	Yes	Yes	Moderate
FUEL EFFICIENCY			
City driving	11.0L/100km	10.6L/100km	10.1L/100km
Highway driving	7.8L/100km	6.9L/100km	6.4L/100km
COST			
4 door	$15,200	$14,300	$11,900
2 door	$13,800	$12,700	$10,600

(Answers, p. 200)

CHAPTER TWELVE

Expository Material

Expository Footnotes and Appendices

When you are writing a report you may want to include some extra information that does not belong in the main body of the text. There are two ways that you can present this information. If it does not take up very much space, you can add it in small print at the bottom of the page, apart from the main text as an *expository footnote*. (Do not confuse *expository footnotes* with *reference footnotes*, which will be discussed in Chapter Fourteen). If there is too much information to fit in a footnote, you can put it in a special section at the end of the report called an *appendix*.

Here are four reasons why you might want to add an expository footnote or appendix.

1. You could have some information related to your topic that the reader might find interesting, but which will not help him understand the topic any better and might even distract him if it appears in the main text. Putting it in an expository footnote or appendix lets him decide whether to read it right away, later, or not at all. A reader who does not want to be distracted can read straight through the report without interruption. Anyone who is interested in knowing more about the subject can take the time to read the footnote or appendix.

   Historical information, for example, is often given in an expository footnote. In a report on detonators that use lead azide, you might mention that mercury fulminate was commonly used before lead azide. Then, in a footnote, you could add:

In 1864 Nobel patented the use of mercury fulminate as a detonator. It was used in combination with potassium chlorate and antimony sulphate.

The reader might find this information interesting, but it will not help him understand the report.

2. The reader may need to be familiar with a particular formula or equation in order to understand your report. Since most of your readers will be experts in the field, they will already know the equation, so it does not need to be included in the main body of the report. However, there may be a few readers who are not familiar with it. For their sake, you could include it in an expository footnote or appendix.

   For example, in a report on electrical potential fields the author might refer to Poisson's equation. If he were writing for electrical engineers he would expect them to be familiar with the equation. However, for those readers who were not, he would add a footnote.

Poisson's equation: $\nabla^2 V = \frac{-\rho}{\varepsilon}$

3. You may want to include a glossary of specific terms that you used in your report. If you are using technical words that your reader might not know, you can add a glossary as an appendix. If you are sure that all your readers will be experts in the field you are writing about, you only need to define terms that are not usually used in that field. However, if you are likely to have readers from other fields, you should include a glossary of all the terms that might need explaining.

   Including definitions of terms is especially important for words that can have two different meanings: a technical one and a common one. For example, to most people the word "horizon" means the line where the earth appears to meet the sky. In soil mechanics, however, it refers to a layer of soil. If you were writing for readers who were not familiar with the terminology of soil mechanics you would have to define "horizon" in terms of its technical meaning.

4. You may want to add a table or graph containing extra information that might interest some readers. If the information in the table or graph is essential to understand your topic, it must be included in the body of the report. It should only be located in an appendix if it is non-essential information.

In summary, *expository footnotes and appendices include information that is supplementary to the text. It is information that most readers will not need in order to understand the text, but which they may find interesting or helpful.* Removing the footnotes or appendices from a report should not affect the reader's ability to understand the text.

EXERCISE ONE

In each of the following examples you will read a thesis statement for a report, followed by several pieces of information related to the topic. In each case, one or two of the pieces of information belong in a footnote or an appendix. Identify that information and explain why it belongs in a footnote or appendix.

1. Thesis statement:
   The purpose of this report is to compare the use of drum brakes to disc brakes in automobiles.
   A. Small wheels, large section tires, and the use of wheel fairings make it difficult to accommodate drum brakes.
   B. The disc brake has a higher resistance to fade than the drum brake.
   C. A predecessor of the modern disc brake was designed for use in jet aircraft in 1945.
   D. Pedal travel with a drum brake increases as the brake heats up.

2. Thesis statement:
   This paper will discuss the problems of chokage, fire and corrosion in tubular air heaters.

A. Chokage is a severe problem with tubular heaters.
B. Chokage is not a problem in regenerative air heaters.
C. Fires are caused by the presence of deposits which build up on the heater surfaces.

3. Thesis statement:
The aim of this report is to show the potential for developing a copper mine on the Sterling property.
A. The initial discovery of copper on the Sterling property was made with a VLF system.
B. Using the VLF system we identified a 600 meter long anomaly on the south west perimeter of the property. This was later identified as a significant copper deposit.
C. VLF stands for very low frequency.
D. VLF is used by the military for submarine communication.

(Answers, p. 201)

If the extra information can be written in one or two sentences, you should usually include it in an expository footnote. An exception to this arises when there are a number of similar extra entries in the report. In that case you might choose to put all the related notes together in one appendix. For example, if you wanted to add several mathematical formulas to the report, you could create an appendix that listed all the formulas together, rather than including each one as a separate footnote.

To create an expository footnote, insert a number right after the word or sentence in the text to which that footnote is directly related. If there is more than one per page, the first footnote will be followed by a number 1, the second by a number 2, and so on. The numbers are written as superscripts—small numbers slightly above the line. The actual footnotes are written at the bottom of the page, beside their matching numbers. Figure 1 on the next page, showing only the bottom part of one page of a report, illustrates how this is done.

...result in more tropical storms, and hence more flooding. Coastal areas, river regions, and the highly erodible soils in the upland areas[1] will all be affected.

On the other hand, in semi-arid zones the increased temperatures will have the reverse effect. In these regions

[1] In the upland areas, damage from flooding has resulted in a 20% decrease in overall productivity in the agricultural sector in the past decade.

Figure 1. Example of an expository footnote

If the information is too long to include in a footnote, it should go in an appendix at the end of the report. A report may have one, several, or no appendices. Each appendix will contain one specific type of information, such as a glossary or a list of proofs or mathematical equations. If there are two or more appendices they are usually labelled alphabetically. For example, the first appendix will be Appendix A, the second will be Appendix B, and so on. Each appendix should have a title indicating what type of information it contains, as in this example.

Appendix A—Glossary
Appendix B—Tables
Appendix C—Proof of Hagen's Theorem

To tell the reader there is an expository footnote you insert a superscripted number in the text. To tell him related information is available in an appendix, you insert a note in brackets right after the sentence or paragraph to which the contents of the appendix are related, as in (See Appendix B).

Graphs, Tables and Diagrams

Sometimes information can be presented more effectively in graphs or tables than in writing. These formats are especially useful if you have to present a lot of numbers or statistics. They are also useful for comparing facts. They present the material in a very organized, easy-to-read way, and usually take up much less space than the equivalent information would take up in paragraph form. For a very simple example, consider Table 1 (below) which shows part of a temperature conversion table to convert from Fahrenheit to Celsius.

The figures in the second and fifth columns show the temperature, in either Fahrenheit or Celsius, that is to be converted. For example, 0°C = 32.0°F, whereas 0°F = -17.8°C.

°C	°C/F	°F	°C°C/F	°F	
-17.8	0	32.0	-12.2	10	50.0
-17.2	1	33.8	-11.7	11	51.8
-16.7	2	35.6	-11.1	12	53.6
-16.1	3	37.4	-10.6	13	55.4
-15.6	4	39.2	-10.0	14	57.2
-15.1	5	41.0	-9.4	15	59.0
-14.5	6	42.8	-8.9	16	60.8
-14.0	7	44.6	-8.3	17	62.6
-13.4	8	46.4	-7.8	18	64.4
-12.9	9	48.2	-7.2	19	66.2

Table 1. Temperature conversion

Now, look at the same information presented in paragraph form.

0°F is equal to -17.8°C. 0°C is equal to 32°F. 1°F is equal to -17.2°C. 1°C is equal to 33.8°F. 2°F is equal to -16.7°C. 2°C is equal to 35.6°F. 3°F is equal to -16.1°C. 3°C is equal to 37.4°F. 4°F is equal to -15.6°C. 4°C is equal to 39.2°F. 5°F is equal to -15.1°C. 5°C is equal to 41°F. 6°F is equal to -14.5°C. 6°C is equal to 42.8°F. 7°F is equal to -14.0°C. 7°C is equal to 44.6°F. 8°F is equal to -13.4°C. 8°C is equal to 46.4°F. 9°F is equal to -12.9°C. 9°C is equal to 48.2°F. 10°F is equal to -12.2°C. 10°C is equal to 50°F. 11°F is equal to -11.7°C. 11°C is equal to 51.8°F. 12°F is equal to -11.1°C. 12°C is equal to 53.6°F. 13°F is equal to -10.6°C. 13°C is equal to 55.4°F. 14°F is equal to -10.0°C. 14°C is equal to 57.2°F. 15°F is equal to -9.4°C. 15°C is equal to 59°F. 16°F is equal to -8.9°C. 16°C is equal to 60.8°F. 17°F is equal to -8.3°C. 17°C is equal to 62.6°F. 18°F is equal to -7.8°C. 18°C is equal to 64.4°F. 19°F is equal to -7.2°C. 19°C is equal to 66.2°F.

This format is much longer than the table and much harder to follow. To prove this, try to use both. Use the chart to find out what 12°C is equal to in Fahrenheit. Now use the 'paragraph' to find out how much 13°F is equal to in Celsius.

Graphs can also be useful. For example, if you are dealing with percentages, a pie graph will give the reader a clear picture of the relevant proportions. The paragraph and pie graph that follow give the same information. Which do you prefer?

There were four major sources of electricity used in Wesland last year. These sources were coal, nuclear power, petroleum, and hydroelectric power. Out of a total of 1,151 billion kilowatt-hours 694 billion kilowatt-hours were provided by coal, 213 billion kilowatt-hours were provided by nuclear power, 143 billion kilowatt-hours were provided by petroleum, 94 billion kilowatt-hours were provided by hydroelectric power and the remaining 7 billion kilowatt-hours were provided by other sources.

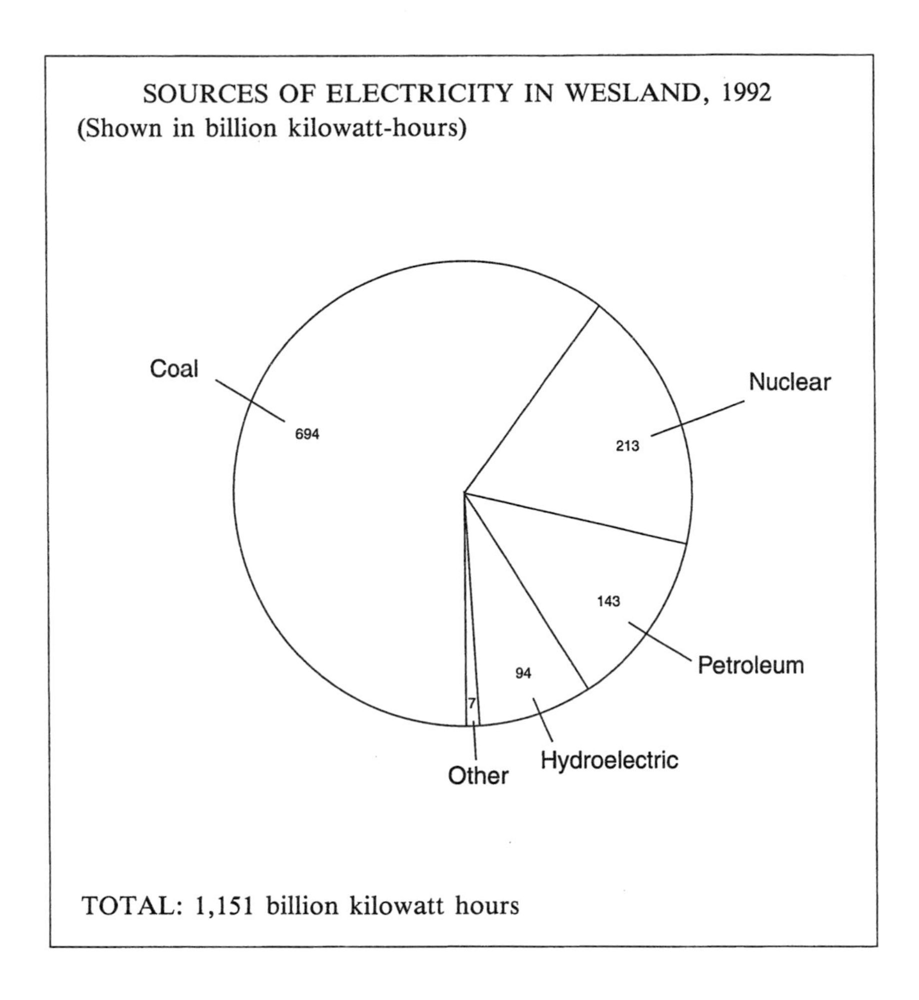

If you include a table, graph or other type of diagram in a report, you must refer to it at least once in the text. If you do not, the reader may not understand its significance or may ignore the information it presents. Also, the information must be relevant to the report and helpful to the reader. You should not include a graph or chart just for the sake of appearance.

EXERCISE TWO

The following data is obtained by making measurements of the earth's magnetic field at different points along a survey line. It would be much easier to follow if it were in table or graph form.

1. Put the data into table form. The table should have two columns. The heading for one column should be "Distance from Origin". The heading for the other column should be "Total Field Magnetic Amplitude".
2. Take the same data and put it on a line graph, with an x and a y axis.
3. What advantage does the table have over the graph? What advantage does the graph have over the table?

DATA

At 0° the total field magnetic amplitude equals 58106. At 25°N the total field magnetic amplitude equals 58111. At 50°N the total field magnetic amplitude equals 58104. At 75°N the total field magnetic amplitude equals 58073. At 100°N the total field magnetic amplitude equals 58011. At 125°N the total field magnetic amplitude equals 57766. At 150°N the total field magnetic amplitude equals 56191. At 175°N the total field magnetic amplitude equals 56206. At 200°N the total field magnetic amplitude equals 59176. At 225°N the total field magnetic amplitude equals 59555. At 250°N the total field magnetic amplitude equals 59115. At 275°N the total field magnetic amplitude equals 58779. At 300°N the total field magnetic amplitude equals 58559. At 325°N the total field magnetic amplitude equals 58284. At 350°N the total field magnetic amplitude equals 58200. At 375°N the total field magnetic amplitude equals 58112. At 400°N the total field magnetic amplitude equals 58130.

(Answers, p. 201)

Titles

All graphs, tables, and other diagrams should have titles. Diagrams and graphs are listed as figures. Tables are listed as tables. For example, the following is the title of a diagram that is included in a report as part of a description:

Figure 1. Lever-type switch

Obviously, Figure 1 would show an illustration of a lever-type switch.

An example of the title of a table is:

Table 1. Estimated run-off of major rivers in Canada

This table would likely have two columns of data. One would list the names of major rivers in Canada; the other would list the amount of run-off for each river.

The first figure in a report will be Figure 1, the second will be Figure 2, and so on. Tables would be numbered accordingly too.

CHAPTER THIRTEEN

Secondary Features

Cover Page: Contents

When you prepare a report you must write a cover page. *A cover page is the page that goes at the front of the report and gives the reader statistical information about the report.* If the report is printed in a journal the information from the cover page is included at the top of the first page of the report.

Several details may be included on the cover page. The format that you follow may be decided by the individual or organization to whom you submit the report, such as a professor, a journal or your employer. *Six features are most commonly found on a cover page.*

1. The title of the report and the name of the author or company that has produced it.

The one item that you will *always* find on a cover page is the title of the report. You will usually find the name of the author(s) too. In some cases you may find the name of the company that has produced it instead of, or together with the name of the specific author.

2. The address of the company or university with which the author is affiliated.

If you are writing a report as a representative of a company you may be expected to include the company's address, telephone number, telex number, and/or fax number. Students are often expected to include a student number and a course identification number.

3. The author's degrees, professional titles and affiliations.
Students or professors at a university submitting a report to a journal usually include their degrees and/or professional titles after their names. They also name the university and the faculty with which they are affiliated. All report writers may be expected to list their affiliation with relevant professional associations or organizations, such as the American Society of Mechanical Engineers (ASME).

4. The date the report was written.
The month and year in which the report was completed should be included. This lets the reader know how up-to-date the report is.

5. A document number.
For reference purposes, companies often give reports a document number so that the report can be easily located in the future.

6. An abstract (sometimes referred to as a summary).
An abstract, which is a concise summary of the contents of the report, is usually located after the cover page, before the introduction. However, it is sometimes included on the actual cover page.

Cover Page: Appearance

If you start working for a new company, you should look at some of their reports to see the format they typically use. Doing so will help you with the overall layout of the report and show you what should be included on the cover page.

The cover page is the very first part of the report the reader sees. Its appearance can influence her attitude toward the report as a whole. If it is untidy or poorly organized, she may react negatively to the entire report before she has even read it.

To produce a cover page that makes a good first impression include only the information that should appear on it and do not crowd it all together in one small space. Do not include information that appears elsewhere such as an abstract that is included on an inside page.

The typeface you use also affects appearance. Vary it if you can, using large print for important items like the title. Less significant details, such as the company's telex and fax numbers, should be in smaller print.

The physical layout of separate items on the page should also reflect the order of importance. For example, the title should be centred in the top half of the page where it will catch the reader's attention. There should be blank space around it to make it stand out. The author's name should be centred below the title. Items such as the company's address, which may not be of immediate interest, should occupy a less prominent space on the page. As well, pieces of information that are related to each other should be in the same general location on the page. For example, the company's address and telephone number should be together.

Figure 1 (on page 147) shows a well designed cover page.

The Ersco ten-cylinder turbocharged marine diesel engine, for use in a 40,000 tonne tanker, is contracted for an output of 21,700 i.h.p. at 115 r.p.m. and an indicated mean pressure of 8.8 kg per sq cm. Actual mean pressure is 10 kg per sq. cm. Bore is 850 mm; stroke is 1,700 mm.

A REPORT ON
THE TEN-CYLINDER TURBOCHARGED
MARINE DIESEL ENGINE

by

Sivakala Prakash

Belco Engineering Research Co.
9545 Waterford Ave
Weston, B.C.
V2X 5T9

April 1984

Tel: (604) 384-8912
Fax: (604) 384-8806

Figure 1. Sample cover page

EXERCISE ONE

Figure 2 shows the cover page for a report on the feasibility of mining the Orgon Deposit in Saskatchewan, written by O. T. Muerthi. Using what you have learned in this chapter state at least five changes that could be made to improve this cover page.

A FEASIBILITY STUDY OF
MINING THE ORGON DEPOSIT IN SASKATCHEWAN

Mazco Ltd.

Doc.# B036192

by O.T. Muerthi

P.O. Box 659
Toronto, Ontario

I would like to acknowledge the assistance of J.P. Fiscott, without whose co-operation this report would never have been completed.

(416)623-9548

Figure 2. A poorly laid-out cover page

(Answers, p. 202)

Title

You must choose the title of your report carefully. Just as the cover page is the first part of the report the reader sees, the title is the first part she reads. It must be brief and concise, but informative too. It should tell the reader exactly what the report is about. If the report is very general, the title should be general. If the report deals with a specific issue, the title should be very specific.

How helpful would "*The Textile Factory*" be as a title? What aspect of the textile factory would a report with this title focus on? Would it be about how a textile factory is designed, about textile factories in general, about the manufacturing processes in a particular factory, or about how that factory was constructed? A title must be much more specific than this one.

Here are three more sample titles:

A. *The History of Textile Manufacturing*
B. *Waste Products of a Textile Factory*
C. *The Construction of the Saba Minch Textile Factory*

They are more informative than "*The Textile Factory*" but they are still vague. Will the report for title A discuss the entire history of textile manufacturing, from the time when ancient man first created cloth? Does title B come from a report describing the potential waste products of a textile factory or from one that deals with ways to dispose of the waste products? Does title C belong to a report about the completion of the Saba Minch Textile Factory, about the work in progress or only about a proposal to construct it? A good title would not leave the reader with so many questions. The following examples are much clearer:

D.	*A Concise History of Automated Textile Manufacturing in Western Europe*
E.	*Cost-efficient Ways to Dispose of Effluent Produced in the Manufacture of Textiles*
F.	*A Progress Report on the Construction of the Saba Minch Textile Factory*

These titles are much more informative. They tell the reader exactly what subjects will be covered, and they do it in only a few words. As a general rule, a title should not be more than one sentence long; however, there are exceptions. Sometimes you may want to give your report a two part title, such as:

GLOBAL WARMING TRENDS: The Effects of Industrial Pollutants on the Environment

The first part of this title states what the broad topic is, and the second part states the specific aspect of that broad topic that the report will cover. Sometimes this style is used when it is difficult to write a good title in one sentence. Other times it is used simply because a writer prefers it. For example, title D (above) could be rewritten as:

AUTOMATED TEXTILE MANUFACTURING IN WESTERN EUROPE: A Concise History

Whichever style you use, you must also consider the layout of the title. It should be centered from left to right and positioned about one third of the way down from the top on the cover page. Sometimes the whole title is printed in capital letters. If it is not, the first letter of each word should be capitalized. However, articles like *the, a,* and *an*, and prepositions such as *of, in, on,* and *to* are not usually capitalized, unless they are the very first word of the title, as in "*T*he History *o*f *t*he ..." or "*I*n Times *o*f...."

One last point to be made about titles is that they are not considered part of the text. This means that the first time you refer in the text to something from the title, you must write it out in full. For example, if you were writing a report called *Insect Problems at the Saba Minch State Farm*, your first sentence in the text could not be, "*The* farm has a serious insect problem." The readers might wonder, "What farm?" You may think it is obvious that you mean the Saba Minch State Farm, but since you have not yet mentioned that in the text, you cannot refer to it that way. Instead, you must say, "The Saba Minch State Farm has a serious insect problem." After that, you could simply refer to "the farm".

EXERCISE TWO

In each of the following examples you will find a brief description of the contents of a report plus three sample titles. Choose the best title.

1. The report analyzes consciousness from a physiological perspective.
   Title A: *CONSCIOUSNESS: A Physiological Analysis*
   Title B: *A Study of Consciousness*
   Title C: *THE QUESTION OF CONSCIOUSNESS: What is it?*

2. The report suggests methods for applying organic chemistry to the treatment of sewage.
   Title A: *Organic Chemistry and Sewage Treatment*
   Title B: *SEWAGE TREATMENT: Suggested Solutions*
   Title C: *Applications of Organic Chemistry to Sewage Treatment*

3. The report outlines the history of synthetic fibres. It includes discussions of the development of viscose, acetate, nylon, terylene, polyethylene, and acrylonitrile.
   Title A: *Viscose, Acetate, Nylon, Terylene, Polyethylene and Acrylonitrile*

Title B: *A History of Synthetic Fibres from Viscose to Acrylonitrile*

Title C: *SYNTHETIC FIBRES: A History*

4. This report compares two methods of preventing corrosion: cathodic protection and protective barriers.

   Title A: *Cathodic Protection and Protective Barriers Against Corrosion*

   Title B: *PREVENTING CORROSION: Cathodic Protection vs. Protective Barriers*

   Title C: *Two Methods of Preventing Corrosion*

(Answers, p. 203)

Table of Contents

A table of contents is a list of the sections in a report. It serves two purposes. It tells the reader what she can expect to find in the report, and it helps her find a specific section quickly.

The best way to explain how to create a table of contents is to discuss an example. Figure 3 (on page 153) is the table of contents for a report about the practice of irrigation in Ethiopia.

CONTENTS

Figure 3. Sample table of contents

One of the first things to learn from this example is that the contents must be listed in the order in which they appear in the report. A table of contents should also clearly indicate which sections are coordinate and which sections are subordinate to each other. In other words, you should be able to tell from looking at the table which sections are sub-sections of other sections. Using numerical notation, as was done in Figure 3 (above), can make these distinctions clear. However, there are other ways to do this, as illustrated in Figure 4.

CONTENTS

Figure 4. Table of contents without numerical notation

The relationships of the sections are still quite clear, thanks to effective use of indentation and of capital and lower case letters. Other features that can be used are **boldface** and different type size and font style. The exact features you use are not important. What is important is that the features reflect the hierarchy of the sections and that you use them consistently.

As well as making layout decisions, you must also decide what to include in the table of contents. If you examine several such tables in a number of different books and reports, you will notice that some are more detailed than others. For example, many authors list only the main headings and omit subheadings. Figures, graphs and tables are often omitted too, but occasionally these are listed in a separate table below the main table. A short report may not even have a table of contents.

The company you work for will probably have a standard format you should follow. You can also look at the tables of contents when you are reading books and reports and see what you find helpful and what you find confusing. As usual, what is important when you are preparing a table of contents is to be consistent. If you list one third level heading, you must list all the third level headings. If you list one figure or table, you must list all the figures or tables.

EXERCISE THREE

The following table of contents belongs to a report that classifies and compares shells. It is very poorly laid out. What changes should you make to improve it? In order to do this exercise you need to know that every order of shells is broken down into superfamilies; in other words a superfamily is subordinate to an order. The page numbers will also give you clues as to how to reorganize it. They should appear in numerical order.

CONTENTS

(Answer, p. 203)

Paging

It is important to number the pages of a report. Without page numbers a table of contents would be useless, and readers would have difficulty finding and keeping track of information.

You may have noticed that the first page listed in the tables of contents in Figures 3 and 4 was not numbered as *2*, but as *ii*. The pages containing secondary features at the beginning of the report are usually numbered with lower case roman numerals like *ii* instead of *II* and *iv* instead of *IV*. The table of contents would be on page i, the abstract and nomenclature on page ii. Page 1 is the first page of the main body of the text, starting with the introduction. This type of numeral—2, 3, 4—is used throughout the rest of the report.

Nomenclature

When you are writing a technical report you might use symbols and/or abbreviations. Even if you think your readers will be familiar with them, it cannot hurt to include a list of their meanings in your report so that uncertain readers can easily look them up.

Such a list may be given the heading "Nomenclature" or "Notation". It is often located at the beginning of the report, before the introduction, but can be put elsewhere. Figure 5 (below) gives an example of this kind of list. It belongs to a report on the cutting performance of silicon carbide alumina.

Nomenclature

B = flank wear length as measured on the rake face (mm)
d = depth of cut (mm)
D = diameter of workpiece (mm)
f = feed rate (mm/rev)
F_C = cutting force (N)
F_F = feed force (N)
FR = radial force (N)
R_a = arithmetic average roughness height (**u w left stick m*)
V = cutting speed (m/min)
VB_B = width of flank wear (mm)
VB_C = width of trailing-edge wear (mm)
VB_N = width of notch wear (mm) Fig.

Figure 5. Sample nomenclature

CHAPTER FOURTEEN

Acknowledgements and References

Acknowledgements

The purpose of the acknowledgements is to formally recognize or show appreciation for any help you have been given in the writing of your report. A secondary function of the acknowledgement section is that it tells the reader where he may be able to obtain information or assistance with work on the same topic.

Many forms of assistance can be recognized or thanked. For instance, you would want to acknowledge an individual or an organization that contributed financially to a project. You might want to acknowledge someone who has given you editorial assistance such as proofreading or typesetting the final report in his free time. You should also acknowledge someone for giving you a useful suggestion or specific information. As well, if you had a research assistant who was not mentioned elsewhere in the report, you might recognize his efforts in the acknowledgements. People sometimes thank their spouses or families too if a project has interfered significantly with their home life, but some firms prefer to omit most personal references.

It is not usually necessary to acknowledge the help of someone within your company who is only doing his job. For example, if a secretary types your report during working hours you do not need to thank him in the acknowledgements.

The "Acknowledgements" section is usually located at the end of the report, after the conclusions but before the references. It may include only one person or organization or it may list several. If a person or organization has made a major contribution, that acknowledgement should be near the top of the list. For example, an organization funding your entire project would come before another organization that gave you one small bit of information. Usually the form the contribution took is listed too. The following is an example of an acknowledgement section:

> *Acknowledgements*
> I would like to acknowledge the financial support of the Northern Geological Society of Bortland without which this project could not have been undertaken.
> Thanks to Professor Adam Bassano for his time and expertise, which contributed greatly to the success of this project.
> Thanks also to Mahnosh Naghibi, Rod Dunstan and Terry Mosler for their editorial assistance.

In this example the author used the phrases:

"I would like to acknowledge...."
"Thanks to...."
"Thanks also to...."

There are many other appropriate variations. Looking at the acknowledgements in several books and reports will help you decide on the style that you would like to follow.

EXERCISE ONE

In each example below, you are told about a report that is already written. Below that, you are given a list of people and organizations that provided the report writers with assistance. Only some of that assistance should be mentioned in the report's acknowledgements. In each case, select the sources that should be included, and write the acknowledgements for that report.

1. Jack works for Chessington Industries. He has written a report about the advantages of using machines in place of human workers in his company's factory. He has received help from the following sources:

Amis Tenant, who works in the same department as Jack, provided Jack with some statistics he (Amis) had obtained on machinery maintenance costs through a study he had done previously.

Ron Briggs, the Vice-President of Tabor Manufacturers Inc., spent one hour with Jack discussing the results of making similar changes at one of his factories.

Jack's secretary typed the report.

A junior clerk at Chessington Industries photocopied and collated the report.

The manufacturer of the machinery that would be used in the factory gave Jack a demonstration of its efficiency.

Jack's wife proofread the report.

2. Joan Borutski, a professor in the Electrical Engineering Department at Biscane University, has written a report about some research she has done on developing long-lasting, recyclable, nickel-cadmium batteries. She has received help from these sources:

   The Electrical Engineering Department of Biscane University gave Joan a substantial research grant.

   Jim MacGilvray, a professor in the Physics Department of Biscane University, gave her some useful suggestions.

   David Baxter, a student, carried out some tests for Joan.

   Dinah Lambden, a lab technician, worked many extra hours assisting Joan.

(Answers, p. 204)

Footnotes and References: When To Include Them

When you write a report you may get some of your ideas from other sources. If you do, some of the material included in your report will not be your own. That is perfectly acceptable as long as you do not take credit for it or suggest that it is yours. To make sure the reader knows whose work or information it is, you add a reference footnote at the point where you refer to it in the text, and you also include a list of your sources in the "References" section at the end of the report.

Professional standards and honesty require that you include these footnotes and references, but they are useful to the reader too. He can refer to them to see where he can find more information on a topic.

Footnotes and References: How to Include Them

There are several possible ways to indicate your sources. Currently, two ways are commonly used in technical journals.

The first involves numbering your sources in the text and then listing them in numerical order in the references at the end of the report. The first time you included information from an outside source such as a report by Joseph Van Emden, for example, you would insert a small number [1] at the end of that information. If a reader wanted to know the source of that material, he would turn to the list of references at the end of the text and look up number 1. There he would find the author's name—Joseph Van Emden, the title of the report, and where, when, and by whom it was published. He would then be able to locate the source if he wanted to do so. The second source in the paper would be number [2], the third would be number [3], and so on.

The second method is to insert the author's last name and the year of publication in brackets right in the text. For example, after using some information from Van Emden's report, published in 1987, you would insert (Van Emden, 1987) in the report. If a reader wanted all the details he could turn to the references at the back and look up Van Emden. When this style is used, the references are listed alphabetically to make it easier to find a particular name.

A variation on this second method involves working the name of the source into the text and listing the year that the work was done or that the source was published in brackets beside the name.

"Van Emden (1987) states that ... " is an example of this variation.

There are advantages and disadvantages to both methods. Using the first approach does not interrupt the text with a name and date. On the other hand, using the second lets the reader see immediately the source's name and how recent the information is. The second method may be easier to write. Using the first method, the author has to keep track of the numbers he uses and also make sure that they correspond to the numbered references at the back. Using the second, he can name the sources right in the text without having to match them up with numbers later.

In the end, however, the method you use may be decided by the company you work for or the publisher of the journal to which you submit the report. Therefore, you should familiarize yourself with both approaches, and with any other variation you may be expected to use.

As mentioned earlier, the references at the back will be listed numerically if you are using the first type of reference footnote, or alphabetically if you are using the second. Whichever method you use, the information that you include for each individual reference will be the same.

The following example shows how a reference to a book is listed following the guidelines of the American Psychological Association. Notice the placement of periods, commas and colons.

> Kahwadjian, A. J. (1986). Gear design for noise reduction (2nd ed.).
> Toronto: Weston Press Ltd.

1. Author's surname and initials.

This should always be the first item in a reference.

> Kahwadjian, A. J.

2. Year of publication in brackets.
Make sure that you put the year of publication of the latest edition, not the latest reprint. If it is not the first edition, this fact should be mentioned in brackets after the title of the book.

> (1986). Gear design for noise reduction (2nd ed.).

3. Title of the book.
Capitalize the first letter of the first word in the title. Underline the title.

> Gear design for noise reduction.

4. Place of publication.

> Toronto:

5. Name of the publishing company.

> Weston Press Ltd.

A journal article follows a slightly different format, as illustrated here:

> Burgensson, M. F. (1992). Hydrochloric acid and dioxin emissions. Combustion Technology, 10, 369-374.

1. Author's name and initials.

> Burgensson, M. F.

2. Year of publication in brackets.

(1992).

3. Title of the article.

Capitalize the first letter of the first word in the title.

Hydrochloric acid and dioxin emissions.

4. Title of the journal.

Capitalize the first letter of each word in the title except for articles and prepositions. Underline the title.

<u>Combustion Technology</u>,

5. Volume and page numbers.

Underline the volume number, follow it with a comma, then list the page numbers.

<u>10</u>, 369-374.

Some additional information:

1. The author's name should stand out. This can be done by extending it to the left of the margin.

Schlee, J. (1964). New Jersey offshore gravel deposits.
    <u>Pit & Quarry</u>, <u>57</u>, 80-82.

2. If the book or article has editor(s) rather than author(s) then (Ed.) or (Eds.) must be inserted after the name:

> Harrow, T.W. (Ed.).

3. Two or more items by the same author are listed according to the year of publication, starting with the earliest publication.

> Suttwill, J. (1987). Mineral sands: An ongoing success. Engineering and Mining Journal, 188, 44-47.
> Suttwill, J. (1989). Alluvial mining grows in popularity. Engineering and Mining Journal, 190, 42-47.

EXERCISE TWO

Use the following information to write a list of references. List them in alphabetical order.

Title:	Fraud and Computers
Authors:	Lamb, A. D. & Beckworth, P.
Date:	Second edition 1988
	Reprinted 1990
Publisher:	F.N. Naft
Place:	London

Title:	Computers: Viral Vandalism
Author:	Audrey C. Rooke
Date:	1989
Publisher:	Beale, Tribley & Scott Ltd.
Place:	Toronto

Title:	High Technology Crimes in the Eighties
Author:	Glendon, Eric. H.
Date:	First published 1989
	Second impression 1990
Publisher:	Amstead University Press

Place: U.K.

Title: Information Access
Authors: Lim, C. & Ward T.
Date: First published: 1981
Second edition: 1984
Publisher: Stoker & Wall Inc.
Place: Boston

Article title: Privacy in the Computer Era
Journal title: Access, Vol. 3, pp. 183-187
Author: Gaudet, P.F.
Date: 1990
Publisher: Access
Place: U.K.

(Answer, p. 205)

Only very basic guidelines for writing footnotes and references have been given here. For greater detail refer to a style manual such as the *Publication Manual of the American Psychological Association.*

CHAPTER FIFTEEN

Abstracts

An abstract is a very short summary of a publication. It states precisely and briefly what the text is about, and may also state what conclusions were reached. An abstract is sometimes called a summary. However, this can be confusing because some writers use the word "summary" instead of "conclusions" as the heading for the concluding section of a report. For that reason the word "abstract" is preferable.

There are several reason why readers of reports find abstracts very useful.

1. **Abstracts are used by professionals to keep up to date on new information in their fields.** Recently graduated students are usually up to date on current developments in their area of specialization. However, once they join the workforce, they do not have time to read every new publication. Falling behind in a field like engineering, for instance, can reduce one's effectiveness after only a few years.

   To get around this problem, many professionals read abstracts. In some cases, all the abstracts from a specific field may be published together in a journal of abstracts. By reading through such a journal regularly, professionals can learn about the latest developments. If they read about something that is particularly interesting or relevant to their work, they can get the full text and read it.

2. **Abstracts are used by students and researchers to find relevant material.** Students and researchers are usually busy people. They

do not have time to read everything that has been written about a topic they are researching. By reading abstracts they can quickly discover which texts have the type of information they want.

3. **Abstracts can help the reader to focus on the main point of a text.** If you read an abstract first, you will be better prepared to follow the main points when you read the actual text. Likewise, if you are having some trouble understanding a text you have just read, reading the abstract again may help because it focuses attention on the main points.

4. **Abstracts are used by librarians when cataloguing material.** Part of a librarian's job is to list every publication in the library according to subject. To do this correctly she must know what each text is about. If the text has an abstract then it will only take her a minute or so to find out exactly what the subject is.

Now that you know what an abstract does, you need to know how to write one. What sort of information belongs in an abstract? What should not be included? The answers to these questions depend on the type of abstract you are writing. *There are two kinds of abstracts:*

- descriptive
- informative

Descriptive Abstract

A descriptive abstract is one that describes the text and what it discusses, without giving specific facts and details. It tells the reader what the subject of the text is, and may state how the author has approached the subject matter, but it does not include concrete data.

Informative Abstract

An informative abstract gives the reader more information. It tells her what the subject of the text is by clearly stating the crucial facts and conclusions found in the text.

Read the two following examples. Both are very short abstracts for the same report comparing two processes for desalting sea water. See if you can identify the descriptive abstract and the informative abstract.

> A. This report compares the propane hydrate process and the butane freezing process for desalting sea water. It examines the energy requirements for each process and analyzes the causes of the differences. This involves comparing the operating temperatures of the two processes and looking at the relative efficiency of the two fuels as refrigerants.

> B. The propane hydrate process for desalting sea water is more efficient than the butane freezing process. The energy requirement of the former is 28% lower. One reason is the smaller heat load. Its operating temperature is 2°C higher than the latter. Also, propane is a more efficient refrigerant in this temperature range.

Example A is a descriptive abstract. It describes the contents of the report without giving any details. Readers know that the report compares the propane hydrate process for desalting sea water to the butane freezing process, but the abstract does not give results of that comparison. Example B, the informative abstract, tells readers which of the two methods is more efficient, and why.

A descriptive abstract has enough information for a librarian cataloguing, or for students and researchers looking for materials, but it will not be too useful to most other abstract readers. A professional trying to keep up to date in her field, or someone who has read the entire text and wishes to see a brief summary of it, wants facts and details that only an informative abstract will provide.

The following guidelines apply to both descriptive and informative abstracts:

1. An abstract is usually between 80 and 300 words.

2. The abstract should make the purpose of the text clear. The purpose does not have to be stated in the form of a thesis statement, but it should be clear from the information given.
3. The abstract should mention any major findings and problems, but not minor ones.
4. The abstract should state the type of procedure followed to produce the findings.
5. Because an abstract is so short, the necessary facts must be stated in as few words as possible.
6. When an abstract is included in the text it only needs the title "Abstract". When it is to be published separately from the main text, as in a journal of abstracts, it must have the same title as the text it summarizes.
7. An abstract may be found at the beginning, the end, or even on the cover of the text. It is most often placed at the beginning, before the introduction.

EXERCISE ONE

After reviewing the guidelines for abstract writing, read the following informative abstract for the report comparing the two processes of desalting sea water. Compare it to Example B on page 169. Is it better or worse than Example B? Why?

C. This report compares two different processes for desalting sea water. One is the propane hydrate process. The other is the butane freezing process. The report concludes that the propane hydrate process is more energy efficient. It shows that the energy requirement for the propane hydrate process is 28% lower than that for the butane freezing process. Two reasons are given for this. One of the reasons is that the propane hydrate operates at a temperature that is about 2°C higher, which results in a smaller heat load. The second reason is that propane is a more efficient refrigerant in this temperature range.

(Answers, p. 205)

EXERCISE TWO

The following is a descriptive abstract for this book, *GETTING TECHNICAL: An Introduction to Technical Writing*. Using thc guidelines for abstracts on pages 169 and 170, you will see that this abstract is very badly written. It is much too long, it is repetitive, and it goes into more detail than is necessary. Following those guidelines, rewrite it.

The book called *GETTING TECHNICAL: An Introduction to Technical Writing* is a book that can be used as a textbook or as a reference book for both speakers of English as a second language and native English speakers, to learn about technical writing. The book has three main parts. The first part of the book focuses on the process of technical writing. It has two chapters. The first chapter is about paragraph structure. The second chapter is about technical writing style.

The second part of *GETTING TECHNICAL: An Introduction to Technical Writing* has four chapters. The first chapter, chapter three, explains how to write definitions. Chapter four explains how to write descriptions. Chapter five explains how to write explanations. Chapter six explains how to write instructions.

The third part of *GETTING TECHNICAL: An Introduction to Technical Writing* focuses on report writing. It has nine chapters. The first chapter, chapter seven, outlines the components and organization of a report. Chapter eight tells the reader how to prepare an outline before he writes his report. Chapter nine talks about the purpose and content of introductions. Chapter ten talks about the purpose and content of the core of the report. Chapter eleven talks about the concluding sections. Chapter twelve talks about expository footnotes, appendices, graphs, tables and diagrams. Chapter thirteen talks about secondary features of the report, such as the cover page, title, and table of contents. Chapter fourteen is about acknowledgements and references. Chapter fifteen, the final chapter, talks about abstracts.

Each chapter of the book includes practical exercises that the reader can do to practice the skills that he has learned in the chapter.

According to this book there are four key factors involved in writing well. The first of these factors is organization. The second of these factors is for the writer to know and state his purpose. The third factor is for the writer to remember who he is writing for, and the fourth factor is the use of simple and straightforward language.

(Answer, p. 205)

EXERCISE THREE

On pages 173-181 you will find a sample report titled "A Comparison of Five Popular All-Season Performance Tires". Write an abstract for it.

(Answer, p. 206)

If you want more practice writing abstracts, read some reports on topics with which you are familiar. Make sure these reports have abstracts but do not read them. First, write your own abstract. Then, read the abstract from the report and compare it to your own. They will be written differently, but both abstracts should include roughly the same information.

A COMPARISON OF FIVE POPULAR
ALL-SEASON PERFORMANCE TIRES

by

David Ellis & Rita Stein

Tire & Tread Inc.
2580 Woodfield Blvd.
Toronto, Ont.
M1X 2V5

Tel: (416) 345-6789
Fax: (416) 345-6799

October 1991

Table of Contents

1. INTRODUCTION

Performance tires are passenger vehicle tires that were designed to improve the handling and high-speed capabilities of automobiles. When first developed, they were only used on sports cars, and could not be safely used with winter road conditions. However, in the mid-eighties an all-season performance tire was developed. This tire combined excellent dry-surface handling with good traction in rain and snow.

Now, most of the major tire makers produce all-season performance tires. They have become very popular and their use is no longer limited to sports cars. In fact, the all-season performance tire now represents approximately 60% of the original equipment market and 40% of the replacement market in tire sales for passenger vehicles.[1]

The difference between one make of all-season performance tire and another is found in its construction and its price. In order to assess the real difference it is necessary to see how the different makes react to various manoeuvres under various road conditions. The aim of this report, therefore, is to compare five popular makes of performance tire that are on the market today.

2. TIRES

The size of tire we chose for these tests is 215/60VR—15. (For an explanation of tire sizes see Appendix A.) This size was chosen as it is a common and popular size that can be used on a broad spectrum of vehicles. The chart in Table 1, below, lists the tires and their specifications.

1

Make & Model of Tire	Tread Depth (mm)	Construction[1]		Wear[2]	Retail Price[3]
		Sidewall Plies	Tread Plies		
Beaufort PQR	8.0	2R	2P,2S,2Pm	180	$130
Mascot JKL	9.0	2P	2N,2S,2Pm	210	$160
Rideau FGH	8.0	2P	2N,2P,2S	170	$140
Rockfort TUV	8.5	2P	2N,2S,2Pm	160	$125
Sharvex RST	8.0	2P	2P,2S,2Pm	180	$145

[1] N = Nylon, P = Polyester, Pm = Polyamide, R = Rayon, S = Steel
[2] Based on the Highway Safety Administration's measurement of durability. 100 points is
acceptable. 200 points is excellent.
[3] The retail price is the average from five major tire retailers.

Table 1. Make, model, & specifications of test tires

3. TESTS

The vehicle chosen for testing the tires was a Pluton PT. It handles well and has very durable anti-lock brakes that were not diminished by the repeated braking required for the tests.

The tires were tested, in both wet and dry conditions, for lateral grip, deceleration, and overall handling and manoeuvrability. As these tires are all-season tires we also performed some tests on snow. The tests carried out were as follows.

2

3.1. Lateral Grip

To rate the tires for lateral grip the vehicle was driven over two laps of a skidpad, one lap in each direction. The lateral acceleration was measured and calculated from electronic timing to 0.001 second. The dry test was carried out on a skidpad with a diameter of 100m. The wet test was carried out on a skidpad with a diameter of 240m with a water depth of 1 to 2mm.

3.2. Deceleration

To measure deceleration the vehicles were accelerated to a speed of 110 kph. Then, the brakes were applied with sufficient force to ensure anti-lock control on all four wheels. The test was carried out three times in each case and the average stopping distance was measured. The dry-braking test was carried out on a straightway. The wet-braking test was carried out on a skidpad equipped with a sprinkler system designed to maintain a uniform water depth.

3.3. Overall Handling

To measure overall handling the vehicles were driven through a 300m slalom course with cones spaced at 30m intervals and the average speed was measured. The same course was used for both wet and dry tests.

3.4. Snow Tests

Acceleration, braking and slalom tests were carried out to determine traction in the snow. All the tests were carried out on a 1500m long, 50m wide straightway that was covered with a layer of moderately packed snow. For the acceleration and braking tests each vehicle was driven on four runs 0—65 kph acceleration and 65—0 kph braking. The slalom tests were set up in exactly the same manner as the wet and dry slalom tests.

3

Unfortunately, increasing temperature and sunshine caused the snow conditions to change drastically, which greatly affected the traction. As a result, the numerical ranking of the test results is highly inaccurate. Nevertheless, we were able to broadly rate the overall performance of each tire on snow as average, above average, or below average.

4. RESULTS AND DISCUSSION

The test results are illustrated in Table 2 below. The number in bold print in brackets below each test result shows how that tire's score rates in relation to the other tires for that test. The tire with the best score will have a 1 in brackets, the tire with the worst will have a 5.

4

Make & Model of Tire	Lateral Acceleration Dry (g)[1]	Lateral Acceleration Wet (g)	Braking Dry (metres)	Braking Wet (metres)	Slalom Dry (kph)	Slalom Wet (kph)	Overall Rating[2]	Snow
Beaufort PQR	0.848 (2)	0.545 (1)	56.5 (2)	80 (1)	99 (2)	58.5 (4)	12 (1st)	Avg.
Mascot JKL	0.817 (4)	0.539 (2)	59.5 (5)	81.5 (2)	97 (5)	63.5 (1)	19 (4th)	Above Avg.
Rideau FGH	0.825 (3)	0.486 (3)	55 (1)	85 (3)	98.5 (3)	63 (2)	15 (2nd)	Below Avg.
Rockfort TUV	0.858 (1)	0.484 (4)	57.5 (3)	85.5 (4)	101 (1)	58 (5)	18 (3rd)	Below Avg.
Sharvex RST	0.804 (5)	0.448 (5)	58 (4)	97 (5)	97.5 (4)	59 (3)	26 (5th)	Above Avg.

[1] g = gravitational force

[2] The overall rating represents the total of the figures in brackets for each of the tests. A low score is better than a high score.

Table 2. Test results.

5

In a lateral acceleration test on a dry surface, an adequate score for a performance tire is 0.8. 0.85 is very good, and 0.9 is excellent. On a wet surface 0.45 is adequate, 0.5 is good and 0.6 is excellent. Though none of the tires excelled, the Beaufort scored highly on both tests. The only tire that did not do well was the Sharvex, which had a barely adequate score for the dry test and an even lower score on the wet.

In a dry braking test, deceleration from 110 to 0 kph should occur within 70m. In a wet braking test, it should occur within 100 metres. All the tires tested within the acceptable limits but the Sharvex took a significantly greater distance than all the others in the wet braking test.

For overall handling and manoeuverability, any speed above 90 kpm is acceptable in the dry slalom test, and any speed above 50 kpm is acceptable in the wet slalom test. All the tires reached speeds well above the minimum.

5. CONCLUSIONS

In the overall rating of the tires for the main battery of tests, the Beaufort PQR got the best rating, followed by the Rideau FGH, the Rockfort TUV, the Mascot JKL and the Sharvex RST in that order. However, when you add the results of the snow tests, the Highway Safety Administration (HSA) wear rating, and the price the results are not quite so straightforward.

Taking all the factors into consideration the Beaufort still clearly comes out with the best overall rating, and unless you have a specific requirement it is probably your best choice of the five. However, all of these tires are good quality tires and if you do have very specific requirements, then these test results should enable you to make an informed decision as to which is the best tire to meet your needs.

6

Acknowledgements

Thanks to Pluton Manufacturers Inc. for providing us with the test vehicle.

Thanks also to S.T. Schott & Sons Ltd. for supplying the 20 wheel rims.

References

[1] Wang, J.B. (1990). "BREAKTHROUGH: The all-season performance tire" in Road, River & Rail, Vol 7, p97.

Appendix A—Specifications for Tire Size

The breakdown for tire size specification 215/60VR—15 is as follows:

'215'is the width of the tire in millimetres.

'60'is the aspect ratio. It expresses the ratio of the height of the sidewall to the width as a percentage.

'V'is the speed rating.*

'R'indicates that it is a radial tire.

'15'is the diameter of the wheel in inches.

*In North America there are six official speed ratings. They are:

S—up to 180 km/h
T—up to 190 km/h
U—up to 200 km/h
H—up to 210 km/h
V—up to 240 km/h
Z—over 240 km/h

7

ANSWER KEY

It is possible to say the same thing in a number of ways. As a result, when exercises involve writing, your answers will not be exactly the same as the sample answers given here. Nevertheless, these answers can be used as guidelines.

CHAPTER ONE—The Paragraph

EXERCISE ONE (p. 7)

1. This paragraph will list the materials used in engineering that have the property of being elastic.
2. This paragraph will look at the various special tools needed for electrical work.
3. This paragraph will state the different sorts of stem assemblies that may be involved.
4. This paragraph will highlight the simple causes of dryer breakdown that should be checked before taking the machine apart.
5. This paragraph will explain what can go wrong if one tightens the jaws of a pipe wrench too much.

EXERCISE TWO (p. 11)

1. There are several ways to classify piles.
2. Piles that are classified according to their effect on soil during installation can be divided into two groups.
3. Displacement piles can be broken down into two groups.

EXERCISE THREE (p. 17)

For this exercise, two sample paragraphs have been given for each answer. In each case, some of the suggestions given in the exercise have been incorporated into sample a, some into sample b, and some into both.

1a. A canister vacuum cleaner has a stronger suction than an upright vacuum cleaner. This is because its motor is more powerful than an upright's. It also has detachable parts, making it less cumbersome than the upright for cleaning drapes, window sills and other areas off the floor. On the other hand, an upright vacuum cleaner is more effective for cleaning rugs. It has a spinning beater brush that is very effective for loosening and removing dirt from a rug.

1b. A canister vacuum cleaner has a stronger suction than an upright vacuum cleaner because it has a more powerful motor. Another advantage is that it has detachable parts. These parts make it less cumbersome than the upright for cleaning drapes, window sills and other areas off the floor. However, an upright vacuum cleaner is more effective for cleaning rugs because it has a spinning beater brush. This brush is very effective for loosening and removing dirt from a rug.

2a. A deep well is required when the water table lies more than 25 feet below the topsoil or when the formations below the water table do not yield water readily. For example, shale absorbs water but it does not yield it. Therefore, if the formations below the water table are shale, a deep well will be needed no matter what the depth of the water table is. A shallow well would be unreliable. Drilling deep wells is expensive because professional drilling equipment is required, but the wells are reliable.

2b. A deep well is required when the water table lies more than 25 feet below the topsoil. It is also required if the formations below the water table do not yield water readily. Shale, for example, absorbs water but does not yield it. This means that if the formations below the water table are shale, a deep well will have to be drilled no matter what the depth of the water table is, since a shallow well would be unreliable. Because drilling deep wells requires professional drilling equipment, it is expensive. However, the wells are reliable.

EXERCISE FOUR (p. 18)

Two sample answers have been given for this exercise.

a. All gasoline engines must complete a sequence of four actions to operate properly. Some engines can do this in two strokes, but most engines require four. First, they must suck in a mixture of fuel and air. This mixture must then be squeezed into a small space. Within that space, the mixture is fired and the force of the burning fuel turns a crankshaft. Finally, the gases created are pushed out of the cylinder into the air. This completes the sequence.

b. All gasoline engines must complete a sequence of four actions to operate properly. First, they must suck in a mixture of fuel and air. Second, this mixture must be squeezed into a small space. Third, the mixture is fired and the force of the burning fuel turns a crankshaft. This creates gases that are pushed out of the cylinder into the air, completing the sequence. Some engines can do this in two strokes, but most engines require four.

CHAPTER TWO—Style

EXERCISE ONE (p. 24)

Number 1 was written for a customer with a broken carburetor. It does not go into much detail, but it tells the reader what he needs to know—what a carburetor is and why it is necessary.
Number 2 was written for children. The language and the explanation are very simple and straighforward.
Number 3 was written for students of auto mechanics. The writer goes into far more detail than in the others and explains the functions of the individual systems within a carburetor.

EXERCISE TWO (p. 26)

1. It is imperative to wear protective clothing at all times when working in the laboratory.
2. Hardened glass lenses cost only a few dollars more than regular lenses.
3. It is illegal to drive when intoxicated.

EXERCISE THREE (p. 27)

Plastics do not occur naturally. They must be manufactured. A few have been made by modifying natural substances such as cellulose, but the majority of them are made by a chemical process called polymerization. Polythene is an example of a plastic produced by this method. It is a white waxy solid obtained by polymerizing ethylene gas (C_2H_4). Polythene's characteristics include...

EXERCISE FOUR (p. 30)

1. The sentence with the active verb, 1b, is better because it is easier to understand.
2. The sentence with the passive verb, 2b, is better. Using the active form required the addition of a vague subject, "people", which is not very suitable.

3. The sentence with the active verb, 3a, is better because it is easier to understand.
4. The sentence with the active verb, 4b, is better because it is easier to understand.
5. The sentence with the passive verb, 5a, is better. The problem with 5b is that the subject, "One of the members of our research team," is long and may distract the reader. As the author does not wish to name the specific person involved, it is better to use the passive.

EXERCISE FIVE (p. 32)

There are many possible ways to rewrite the example given. The following is one sample answer.

To remove an outboard motor from a boat you must follow these steps. First, disconnect the fuel line and run the motor in neutral until all excess fuel is removed from the carburetor. Next, disconnect the remote control, electric cable and safety chain from the motor. After that, loosen the thumbscrews and tilt the motor to drain off the water. Finally, lift the motor from the boat.

EXERCISE SIX (p. 34)

There is more than one way to rewrite the paragraph. The following is one sample answer.

The water turbine, a descendant of the waterwheel, was invented to drive machinery in mills and factories. However, it was invented when steam power was coming into heavy use, so it was rarely used until the 1870's.

EXERCISE SEVEN (p. 36)

1. The answer is 1b. 1a is too vague.
2. The answer is 2a. Ford does manufacture station wagons, but it also manufactures many other types of automotive vehicles. Therefore, using the broader category—automotive vehicles—is as specific as you can get.
3. The answer is 3a. 3b is too vague.
4. The answer is 4a. It gives the reader specific details about which tools to use. 4b is a longer sentence, but it is very vague.
5. The answer is 5b. 5a offers more details, but it is inaccurate. It suggests that the tips of all pens have rolling balls when, in fact, only ballpoint pens do.

CHAPTER THREE—Definitions

EXERCISE ONE (p. 43)

1. A magnetic compass is an instrument used to determine direction.
2. A sickle is an implement used for cutting grass.
3. A pair of water pump pliers is a tool that is used for gripping pipes.
4. A universal cutter and tool grinder is a machine that can perform a wide variety of grinding operations.
5. A machinist's vise is a device that is used to hold metal or wood securely in place.
6. A gearshift is a mechanism that changes gears in a transmission.
7. A tripod is a piece of surveying equipment.
8. A silo is a tall structure that is used for storing fodder.

EXERCISE TWO (p. 45)

1. A bus is a vehicle that is used to transport large numbers of people. It resembles a very large car.
2. A microwave oven is an appliance that is used for heating foods quickly. It is found in both home and restaurant kitchens.
3. A sofa is a piece of furniture that is used to seat several people. It looks like a very wide armchair.
4. A bulldozer is a piece of construction equipment that is used for moving dirt. It looks like a tractor with a wide steel blade attached to the front.

CHAPTER FOUR—Descriptions

EXERCISE ONE (p. 51)

1. The surface is 30 cm long x 20 cm wide. The area of the surface is 600 square centimeters **OR** 600 cm^2.
2. The base has a capacity of 1800 cm^3.
3. The steel must be at least 0.2 mm thick.
4. The width of the beam is 11 cm. **OR** The beam is 11 cm wide.
5. The height of the building is 50 m.
6. The volume of the cage is 60 cm x 20 cm x 40 cm (a third dimension is required for volume).
7. The depth of the ditch is two meters.
8. The rod is 2.5 cm in diameter.
9. The weight of the stone is 5 kg. **OR** The stone weighs 5 kg.
10. The track is 7.8 m long.

EXERCISE TWO (p. 54)

1. Both these descriptions give the same information, but B is better. The reason is the order in which the information is given. A is incorrect because the dimensions are given first, followed by a definition. The definition should always come first.
2. Description A is the better one. It tells the reader what soccer balls are made of. This information is important in technical writing. Description B tells the reader what color the soccer ball is. Color only needs to be mentioned in a technical description when items are color-coded. For example, electrical wires or pipes that look very similar may be color-coded to help a worker identify the correct one. However, if the color of an object is not used for identification purposes, it need not be mentioned.
3. Description B is better. In description A the dimensions have been left out. Someone who did not know what a pen was could imagine it to be several times its actual size.

EXERCISE THREE (p. 55)

1. A die is a small cube with a different number of spots, ranging from 1 to 6, on each side. It is used for playing games and gambling. It is usually made of plastic or wood. The width of one side of an average die is between 1 cm and 2 cm.
2. A pencil is a cylindrical writing utensil. It is wood with a graphite core. One end of the pencil is sharpened to a point that is used for writing or drawing. A typical pencil is 18 cm long with a diameter of 7 mm.
3. Jam jars are cylindrical containers that are used for storing fruit preserves. They are usually made of glass, with plastic or metal lids. A typical jam jar is between 8 and 15 cm high, with a diameter ranging from 6 cm to 9 cm.

EXERCISE FOUR (p. 56)

The table is a square, solid oak end table. It has four square legs that are flush with the corners of the table. It is 58 cm long, 58 cm wide, and 45 cm high. The table top is 4 cm thick. The legs measure 8 cm by 8 cm.

EXERCISE FIVE (p. 64)

The following are sample answers only. Your answers will not be exactly the same, but they should include much of the same information.

1. A septic tank is a watertight receptacle that is used to separate solid waste from liquid sewage. The solids are stored until they are sufficiently broken down to be discharged for final disposal. The tank may

have one or two compartments. It must have an inlet device at one end and an outlet device at the other end. Tightly sealed covered manholes should provide access to the inlet and outlet devices.

Septic tanks come in a number of shapes. The most common are horizontal or vertical cylindrical tanks and horizontal rectangular tanks. They are generally made of pre-cast concrete or welded sheet steel. A septic tank for a four bedroom house should have a capacity of at least 4,500 liters.

The inlet device consists of an inlet pipe and a vented tee or baffle that diverts the incoming sewage downwards. The outlet device also consists of a vented tee or baffle and an outlet pipe. In a gravity-feed system the outlet pipe must be several centimeters lower than the inlet pipe.

2. The SR2 is a single compartment septic tank made of pre-cast concrete. It is a rectangular tank with an inlet device consisting of a concrete inlet pipe and a concrete baffle at one end, and an outlet device consisting of a concrete baffle and a concrete outlet pipe at the other end. Tightly sealed covered manholes in the surface of the tank provide access to the inlet and outlet devices.

   The tank is 2.75 m long, 1.25 m wide and 1.5 m high, and it has a capacity of 3,400 liters. The inlet and outlet pipes are 10 cm in diameter. The inlet pipe is situated 18 cm below the surface of the tank, and the outlet pipe is situated 25 cm below the surface. The baffles, which are 5 cm thick and 60 cm high, extend the width of the tank. They are situated 3 cm below the top of the tank at a distance of 18 cm from the end walls. The manhole covers are 50 cm by 50 cm.

CHAPTER FIVE—Explanations

EXERCISE ONE (p. 67)

Sentence 1 is irrelevant. It tells the reader a historical fact. It does not explain how the thermometer works. Sentences 2 and 3 are relevant. They both help to explain how the thermometer works.

Sentence 4 is irrelevant. The thermometer we are writing about is only used to measure temperatures between the range of 33°C and 41°C. The fact that mercury can function at far higher and lower temperatures is interesting but not relevant to an explanation of how this thermometer works.

Sentence 5 is irrelevant. It gives us information about other types of thermometers, but not about the one we are dealing with.

Sentence 6 is relevant. It helps to explain why the thermometer works.

Sentences 7 and 8 are both relevant because they help explain how the thermometer works.

Sentence 9 is irrelevant. It belongs in instructions since it tells the reader how to use the thermometer, but it does not explain how it works.

EXERCISE TWO (p. 69)

(Definition) A ballpoint pen is a cylindrical writing utensil. **(Description)** It consists of a metal or plastic nib attached to one end of a narrow ink-filled tube called a reservoir. The reservoir is encased in a slightly wider plastic tube. The average ballpoint pen is 14 cm long and 7 mm in diameter.

A small ball, which is usually made of steel, is housed in a socket at the tip of the nib. **(Explanation)** When the tip of the pen is pressed down and moved across a piece of paper, the ball rolls. As it rolls, it transfers ink from the ink tube to the paper.

EXERCISE THREE (p. 70)

The following is a sample answer. Yours will probably not be written in exactly the same words.

A mathematical compass is an instrument that is used for drawing circles and arcs. It consists of two arms of approximately equal length. These arms are hinged at one end so that they can be moved together or apart. One of the arms ends in a sharp point while the other arm holds a pencil lead at the end. This instrument may be made of metal or plastic. The arms of a typical compass are 12 cm long.

To draw a circle, the arms are opened until the distance between the two outer points is equal to the radius of the circle. Then the sharp point is placed where the center of the circle will be on a piece of paper. It is held in place while the other arm pivots around it with the lead pressed against the paper. In this way, the circle is marked on the paper.

EXERCISE FOUR (p. 73)

A magnetic compass is an instrument that is used to determine direction. It resembles a small circular container made of plastic or metal with a clear glass or plastic cover. Inside the container there is a magnetized needle and a compass card that indicates the 32 points of direction. A typical hand-held compass has a diameter of 5 to 10 cm and is 1 to 2 cm deep.

The magnetic needle is poised on a point in the center of the compass in such a way that it can pivot freely. No matter which way the compass is held, the needle will always swing round so that the needle is pointing toward the magnetic north pole. By rotating the compass so that north—as indicated on the compass card—is aligned with the needle, you can also see where all the other directions lie.

CHAPTER SIX—Instructions

EXERCISE ONE (p. 77)

1. Pour 400 cm^3 of water into a beaker.
2. Add 80 grams of powdered alum to the water.
3. Place the beaker in a can that is half full of water.
4. Heat the can over a bunsen burner for two to three minutes.

EXERCISE TWO (p. 78)

How to Start a Car

1. Fasten your seat belt as soon as you get in the car.
2. Check the gear shift lever to make sure the car is in neutral gear. If it is not, depress the clutch pedal with your left foot and move the gear shift into neutral position, as shown on the gear shift handle.
3. Make sure the hand brake is on. (The lever must be up, not horizontal).
4. Put the key in the ignition slot. Turn the key clockwise as far as it will go, while pressing on the accelerator pedal with your right foot, until the engine starts. Release the key. If the engine falters, press down again on the accelerator pedal.
5. Move your right foot onto the brake pedal, press it down and hold it there.
6. Release the hand brake.
7. Keeping your right foot on the brake, depress the clutch pedal with your left foot. While it is depressed move the gear shift into first gear, as shown on the gear shift handle. Move your right foot onto the accelerator pedal and slowly push down on it while gently releasing the clutch pedal. The car will start to move forward.

EXERCISE THREE (p. 80)

How to Fix a Flat Tire on a Bicycle

EXERCISE FOUR (p. 81)

Tools and materials required:
Two wrenches
Two tire levers
A bicycle pump
A bucket of water

A piece of chalk or a marker
A small piece of coarse sandpaper
A rubber patch, 2 to 3 cm in diameter
Rubber cement

EXERCISE FIVE (p. 82)

Using two wrenches, loosen the bolts on either side of the wheel hub. Remove the wheel from the frame. Then, using two tire levers, remove the inner tube from the wheel.

Next, pump up the inner tube. Then, place your hands about 15 cm apart on the inner tube, place it in a bucket of water and squeeze. If you see air bubbles rising from one spot, you have located the hole. If not, move your hands along the inner tube and repeat the procedure. Continue moving around the inner tube in this fashion until you have located the hole. Mark it with a piece of chalk or a marker.

Press down in the center of the valve to release the air from the inner tube. Take a small piece of coarse sandpaper and roughen the surface of the inner tube over an area around the hole that is 3 to 4 cm in diameter. Get a rubber patch with a diameter of 2 to 3 cm. Cover one side of the patch with rubber cement. Also cover the roughed up area of the inner tube with rubber cement. Wait approximately 30 seconds until all the cement appears to have evaporated. Then, cover the hole with the patch, cement side down, and press it firmly.

Finally, put the inner tube back on the tire, and put the tire back on the wheel. Put the wheel back on the bicycle, tighten the nuts on the wheel hub, and inflate the tire.

EXERCISE SIX (p. 83)

The following are possible headings:
Removing the inner tube
Locating the puncture
Fixing the puncture
Reassembling the bicycle

EXERCISE SEVEN (p. 83)

How to Fix a Flat Tire on a Bicycle

Tools and materials required:
Two wrenches
Two tire levers
A bicycle pump
A bucket of water

A piece of chalk or a marker
A small piece of coarse sandpaper
A rubber patch, 2 to 3 cm in diameter
Rubber cement

Removing the inner tube

1. Using the wrenches, loosen the bolts on the hub on either side of the wheel.
2. Remove the wheel from the frame.
3. Using the tire levers, remove the inner tube.

Locating the puncture

1. Pump up the inner tube.
2. Place your hands about 15 cm apart on the inner tube, place it in a bucket of water and squeeze. If you see air bubbles rising from one spot you have located the hole. If not, move your hands along the inner tube and repeat the procedure. Continue moving around the inner tube in this fashion until you have located the hole.
3. Mark the hole with a piece of chalk or a marker.

Fixing the puncture

1. Press down in the center of the valve to release the air from the inner tube.
2. Take a small piece of coarse sandpaper and roughen up the inner tube over an area around the hole that is 3 to 4 cm in diameter.
3. Get a rubber patch with a diameter of 2 to 3 cm. Cover one side of the patch with rubber cement. Cover the area of the inner tube around the hole with rubber cement.
4. Wait approximately 30 seconds until all the cement appears to have evaporated.
5. Cover the hole with the patch, cement side down, and press it firmly.

Reassembling the bicycle

1. Put the inner tube back in the tire.
2. Put the tire back on the wheel.
3. Put the wheel back on the bicycle.
4. Inflate the tire.

EXERCISE EIGHT (p. 84)

The following is a sample of more specific instructions:

How to Make a Local Telephone Call

1. Put 25 cents in the slot in the upper right hand corner. You may use any combination of coins.

2. Lift up the receiver.
3. Listen for a dial tone. If there is no dial tone, hang up and try again.
4. Dial the number you wish to call.
5. If the line is busy, hang up the receiver. Your money will be returned in the slot in the lower right hand corner of the box.

EXERCISE NINE (p. 85)

How to Replace the Focusing Screen

1. Detach the lens from the camera body.
2. Grasp the focusing screen release latch at the front of the mirror box casting with tweezers and pull it until the holder springs open.
3. Take hold of the small tab on the screen with the tweezers and lift the screen out of the holder.
4. Take hold of the small tab on the new screen with the tweezers and carefully place it in position in the holder, with the flat side facing downward.
5. Using the tweezers, gently push down on the tab until it clicks into place in the holder.

CHAPTER SEVEN—Layout

EXERCISE ONE (p. 96)

1. INTRODUCTION

2. METHODS OF PREVENTION

2.1 Modifying the Metal's Properties

2.2 Modifying the Environment Around Metals

2.3 Protective Barriers

2.3.1. Electroplating

2.3.2. Painting

2.3.3. Cathodic protection

3. CONCLUSIONS

EXERCISE TWO (p. 98)

1. INTRODUCTION

xxxxxxxxxxxxxxxxxxxxxxxxxxxxxxxxxxxxxxxxxxxxxxxxxxxxxxxxxxxxxxx

2. FERTILIZERS

xxxxxxxxxxxxxxxxxxxxxxxxxxxxxxxxxxxxxxxxxxxxxxxxxxxxxxxxxxxxxx

2.1. Nitrogenous Fertilizers

xxxxxxxxxxxxxxxxxxxxxxxxxxxxxxxxxxxxxxxxxxxxxxxxxxxxxxxxxxxxxx

2.1.1. Natural sources

xxxxxxxxxxxxxxxxxxxxxxxxxxxxxxxxxxxxxxxxxxxxxxxxxxxxxxxxxxxxxx

2.1.2. Manufactured sources

xxxxxxxxxxxxxxxxxxxxxxxxxxxxxxxxxxxxxxxxxxxxxxxxxxxxxxxxxxxxxx

2.2. Phosphatic Fertilizers

xxxxxxxxxxxxxxxxxxxxxxxxxxxxxxxxxxxxxxxxxxxxxxxxxxxxxxxxxxxxxx

2.2.1. Natural sources

xxxxxxxxxxxxxxxxxxxxxxxxxxxxxxxxxxxxxxxxxxxxxxxxxxxxxxxxxxxxxx

2.2.2. Manufactured sources

xxxxxxxxxxxxxxxxxxxxxxxxxxxxxxxxxxxxxxxxxxxxxxxxxxxxxxxxxxxxxx

3. APPLYING FERTILIZERS TO THE SOIL

xxxxxxxxxxxxxxxxxxxxxxxxxxxxxxxxxxxxxxxxxxxxxxxxxxxxxxxxxxxxxx

4. CHEMICALS

xxxxxxxxxxxxxxxxxxxxxxxxxxxxxxxxxxxxxxxxxxxxxxxxxxxxxxxxxxxxxx

4.1. Herbicides

xxxxxxxxxxxxxxxxxxxxxxxxxxxxxxxxxxxxxxxxxxxxxxxxxxxxxxxxxxxxxx

4.2. Insecticides

xxxxxxxxxxxxxxxxxxxxxxxxxxxxxxxxxxxxxxxxxxxxxxxxxxxxxxxxxxxxxx

4.3. Fungicides

xxxxxxxxxxxxxxxxxxxxxxxxxxxxxxxxxxxxxxxxxxxxxxxxxxxxxxxxxxxxxx

5. APPLYING CHEMICALS

xxxxxxxxxxxxxxxxxxxxxxxxxxxxxxxxxxxxxxxxxxxxxxxxxxxxxxxxxxxxxx

6. CONCLUSIONS

xxxxxxxxxxxxxxxxxxxxxxxxxxxxxxxxxxxxxxxxxxxxxxxxxxxxxxxxxxxxxx

EXERCISE THREE (p. 99)

2. Gels

Axle-grease, an emulsion of insoluble metal soap in lubricating oil, is fluid under pressure and fairly solid when not. Suspensions in water of bentonite, a form of fuller's earth, are used to lubricate rock drills because they set when the drill is stationary and move freely otherwise. These are examples of *thixotrophy*.

The particles of the disperse phase are plate-like or rod-like in shape. . . .

3. Foams

Pure liquids do not ordinarily foam. For foams to be created there must be a surface-active agent. . . .

EXERCISE FOUR (p. 102)

1. In section 1 the main heading has a double underline. In section 2 the main heading has only a single underline.
2. In section 1 the smaller headings only have a capital letter at the beginning of the first word. In section 2 each word in the headings starts with a capital letter.
3. Heading 1.1.3. is **Soil composition**. Heading 2.1.3. is **Composition of Soil**. They should be the same.
4. In section 1, none of the heading numbers ends with a period. In section 2 all of the heading numbers end with periods.

CHAPTER EIGHT—Outlines

EXERCISE ONE (p. 109)

A. This statement of purpose states what the main point of the report is, but it does not say for whom it is being written. As well it does not state the writer's intentions—to convince James & Johnson that Baldwin's is not responsible for the delay.

B. This statement of purpose does not mention the main point of the report—to document the progress of the construction. Also, it suggests that the writer's aim is to blame James & Johnson for the delays, rather than just to convince them that Baldwin's is not responsible for the problems.

C. James & Johnson are already aware that the construction will not be completed on time. This statement of purpose does not mention the real purpose of the report—to document the progress. As well, it does not state the writer's intentions.

Sample Statement of Purpose:

The purpose of this report is to document progress on the construction of Glen Lodge for its owners, James & Johnson. The construction is behind schedule. I must convince James & Johnson that Baldwin's is not responsible for the delays and has done everything possible to avoid them.

EXERCISE TWO (p. 109)

This report is being written for a bicycle manufacturing company called Welbourne Cycle. This company is planning to use robots for the purposes of welding and assembly, and has to decide whether to buy two single purpose robots or one general purpose robot.

The report writer will analyze the company's present needs. He will also look at the company's projected future needs. He will then compare the costs of buying, operating and maintaining the different types of robots. He will also compare the efficiency of the different robots. Specifically, he will compare the time it takes to perform certain tasks, the quality of the output, and how efficiently the materials are used.

He will look at the results of these comparisons and recommend that Welbourne Cycle buy either one general purpose or two single purpose robots.

He will get most of his information from Welbourne Cycle, from the companies that manufacture the robots, and from companies that are using the robots.

CHAPTER NINE—Introductions

EXERCISE ONE (p. 113)

1. The aim of this report is to examine the problems related to using DNA for forensic identification. The report will highlight the areas in which further research needs to be carried out before DNA sampling is used for forensic identification purposes.
2. This report will examine several alternatives to the present practice of cleaning coal by scrubbing sulfur dioxide from flue gases.
3. The aim of this report is to explain how cell reproduction is regulated. Specifically, it will focus on recent research which shows that cell reproduction is regulated by the xyz protein.

EXERCISE TWO (p. 114)

(Background Information) With increasing concern over the environment, scientists have been searching for a viable alternative to chemical pesticides. Genetic engineering, which is one such alternative, has received a lot of publicity but has met with only limited success. A lesser known area of research has focused on the use of micro-organisms to destroy pests.

It has been known for fifty years that various types of micro-organisms, called nematodes, are parasitic to insects in the larval stage. Because mass production and transportation of the organisms proved problematic, little research was done in the area until recently. Now, however, scientists have made a break-through that will overcome these problems. (Thesis Statement) The purpose of this paper is to show that with the new technology, nematodes can provide an economically viable and environmentally friendly alternative to many chemical pesticides and other biological alternatives.

(Limiting Factors) Because nematodes live underground, their efficiency is limited to the destruction of pests that spend part of the initial stage of their life cycles underground. Unfortunately only approximately 30% of the insects known to be harmful to agriculture do this. Therefore, it is not the aim of this paper to suggest that nematodes can replace chemical pesticides altogether.

EXERCISE THREE (p. 117)

Factory workers are continually being replaced by automated machinery. In many ways, this machinery is more efficient than the human worker. However, humans have the advantage of being able to learn new tasks, whereas automated machinery designed to perform a specific task cannot easily be adapted to do anything else.

However, with the rapid development of computer technology, this limited adaptability is changing. By combining a design that simulates the movement of a human arm and hand with a computerized control system, scientists have developed a new type of machine that can be programmed to perform an almost infinite variety of tasks. The machine is called a robot.

This report will look at the advantages and disadvantages of using robots in manufacturing. To do this, we will compare the overall productivity and efficiency of robots to that of automated machinery and human workers in a set variety of tasks.

EXERCISE FOUR (p. 118)

Most of the machinery that is used in automatic manufacturing is designed to perform a specific task and cannot be adapted to do anything else. As a result, when a company wants to make any major changes to its manufacturing process, it has to acquire costly new equipment.

About fifteen years ago, a group of electronic and mechanical engineers who were working to overcome this problem developed the robot. A robot is an adaptable machine that can be programmed to perform a great variety of tasks. The purpose of this report is to outline the advantages of using robots in manufacturing.

Although any one robot can perform a wide variety of tasks, the nature of those tasks is limited. It is not, therefore, the author's intention to suggest that all other forms of automatic machinery are obsolete.

CHAPTER TEN—The Body of the Report

EXERCISE ONE (p. 124)

1. Introduction
2. Preliminary Test
   2.1. Friction Oxidation Rig
   2.2. Plain Bearing Oscillating Rig
   2.3. Ball Oscillating Rig
3. Lubricant X
   3.1. Friction Oxidation Rig
   3.2. Plain Bearing Oscillating Rig
   3.3. Ball Oscillating Rig
4. Lubricant Z
   4.1. Friction Oxidation Rig
   4.2. Plain Bearing Oscillating Rig
   4.3. Ball Oscillating Rig
5. Results & Discussion
6. Conclusions

EXERCISE TWO (p. 126)

1. A and C are relevant. B is not. The information in B is related to the topic, but it is not relevant to the report which is only comparing propane to electricity. It is not dealing with other fuels such as oxy-acetylene. (Information of this kind could possibly have a place in the introduction of a report, but it is unlikely to belong in the main body.)

2.

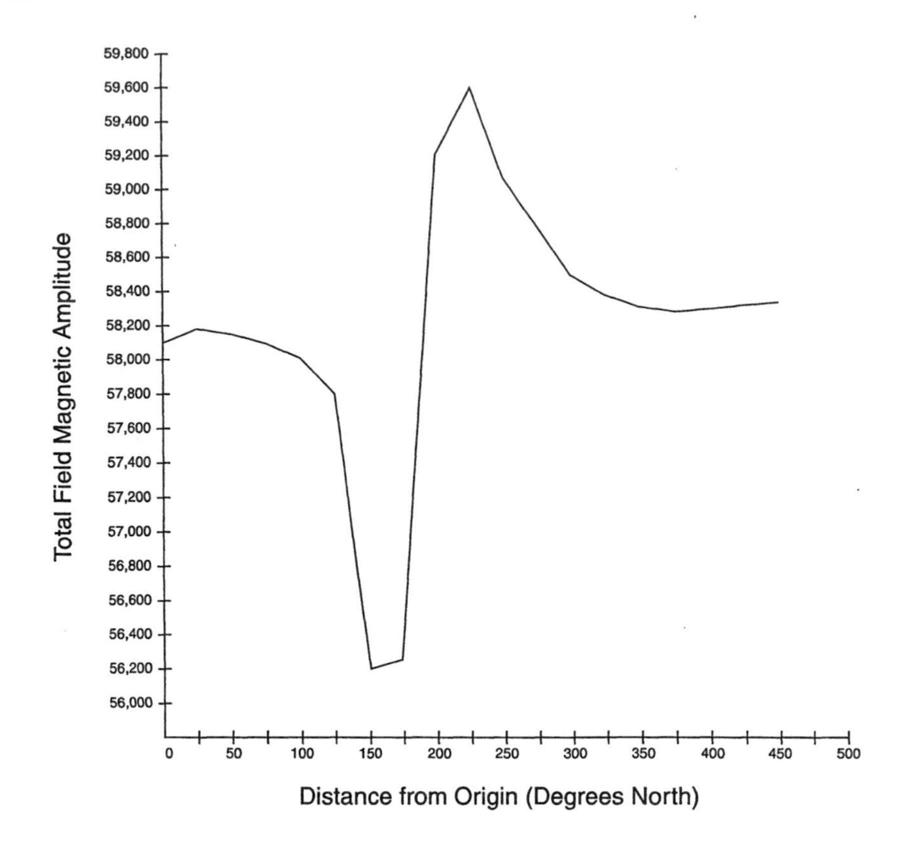

3. It is easier to find the exact figure for the total field magnetic amplitude on the table than it is on the graph. The graph has the advantage of giving readers an immediate visual image of how the measurements fluctuate over distance.

CHAPTER THIRTEEN—Secondary Features

EXERCISE ONE (p. 148)

The following changes should be made to the cover page:

1. The title should be in the center of the page.
2. The title should be in typeface that will stand out.

3. The document number should be in smaller type and should occupy a less prominent position on the page
4. The date is missing and should be added.
5. The writer's name should be centered below the title.
6. The name and address of the company should be together.
7. The acknowledgements should not be on the cover page.

An improved version of the cover page might look like this:

A FEASIBILITY STUDY OF
MINING THE ORGON DEPOSIT
IN SASKATCHEWAN

by
O.T. Muerthi

Doc.# B036192
Mazco Ltd.
P.O. Box 659, Stn. 'Z'
Toronto, M4Q 1C9
Tel: (416) 623-9548

May, 1992

EXERCISE TWO (p. 151)

1. A
2. C
3. B
4. B

EXERCISE THREE (p. 155)

There are several problems with the table of contents:

1. The contents should be listed in the order in which they appear. To put them in the correct order, look at the page numbers and organize them accordingly.
2. You were told that a superfamily is a sub-category of an order. That means that all the orders are major headings and all the superfamilies are subheadings. The layout of the table of contents should make that fact clear, either by using different print (larger, capitalized or bold

print for *all* the major headings) and/or by indenting *all* the superfamilies.

3. Introduction is a major heading. It should be printed in the same type as all the other major headings.
4. None of the headings should start with lower-case (small as opposed to capital) letters. In the example, Introduction and some of the names of the superfamilies start with lower-case letters.

The following example shows how the table could be rewritten:

CONTENTS

Abstract ii

INTRODUCTION 1
ORDER: ARCHAEOGASTROPODA 1
  Superfamily: Pleurotamariacea 2
  Superfamily: Fissurellacea 3
  Superfamily: Patellacea 4
  Superfamily: Trochacea 6
ORDER: MESOGASTROPODA 7
  Superfamily: Littorinacea 7
  Superfamily: Cerithiacea 8
  Superfamily: Epitoniacea 10
ORDER: NEOGASTROPODA 11
  Superfamily: Muricacea 12
  Superfamily: Buccinacea 13
  Superfamily: Volutacea 14
  Superfamily: Conacea 14
SUMMARY 15
References 17
Appendix—Glossary 18

CHAPTER FOURTEEN—Acknowledgements and References

EXERCISE ONE (p. 159)

1. Acknowledgements

   I would like to thank Mr. Ron Briggs, Vice President of Tabor Manufacturers Inc., for giving me his valuable time and information. I would also like to thank my wife for her editorial assistance.

2. Acknowledgements

   I would like to acknowledge the help of the following, without whose help this research would not have been possible:
   David Baxter, who carried out some tests on the project.
   Dinah Lambden, who put in many extra hours in the lab.
   Professor Jim MacGilvray, for giving me some useful suggestions.

EXERCISE TWO (p. 165)

Gaudet, P.F. (1990). Privacy in the computer era. Access, 3, pp. 183—187. U.K.: Access.

Glendon, E.H. (1989). High technology crimes in the eighties. U.K.: Amstead University Press.

Lamb, A.D. & Beckworth, P. (1988). Fraud and computers (2nd. ed.). London: F.N. Naft.

Lim, C. & Ward, T. (1984). Information access (2nd. ed.). Boston: Stoker & Wall Inc.

Rooke, A.C. (1989). Computers: viral vandalism. Toronto: Beale, Tribley & Scott Ltd.

CHAPTER FIFTEEN—Abstracts

EXERCISE ONE (p. 170)

The first abstract (B) was better than the second one (C) because of the style in which it was written. It gave the reader exactly the same information using fewer words. (C) would have made a better paragraph in the body of the actual text because it is more cohesive, but in an abstract it is more important to be brief.

EXERCISE TWO (p. 171)

There is no one correct way to rewrite this abstract. The following is a sample answer.

GETTING TECHNICAL: An Introduction to Technical Writing is a textbook/reference book about technical writing. Written both for speakers of English as a second language and native English speakers, it has three parts. The first part focuses paragraph structure and style. The second part explains how to write definitions, descriptions, explanations and instructions. The third part focuses on report writing, outlining the purpose and content of the component parts of a report. Each chapter also includes practical exercises.

EXERCISE THREE (p. 172)

Abstract

Five all-season performance tires, size 215/60VR—15, including the Beaufort PQR, the Mascot JKL, the Rideau FGH, the Rockfort TUV and the Sharvex RST, were tested in wet and dry conditions for lateral grip, deceleration and overall handling. Snow tests were also carried out to determine traction. The Beaufort PQR got the best overall rating. The Sharvex RST got the lowest rating even though it did excel in the snow.